THE ORIGINAL AND
GROWTH OF PRINTING
Together with
THE KING'S
GRANT OF PRIVILEGE
FOR SOLE PRINTING
COMMON-LAW-BOOKS
DEFENDED
and
THE VINDICATION OF
RICHARD ATKYNS

Frontispiece from the 1664 edition of *The Original and Growth of Printing*.
Image © British Library Board, shelfmark 129.a.7.[1]

THE
Original and Growth
OF
PRINTING

Together with

THE KING'S GRANT
OF PRIVILEGE FOR
SOLE PRINTING
COMMON-LAW-BOOKS
DEFENDED

and

THE VINDICATION OF
RICHARD ATKYNS ESQUIRE

𝕽𝖎𝖈𝖍𝖒𝖔𝖓𝖉
TIGER OF THE STRIPE
MMXIII

This edition first published in 2013 by
Tiger of the Stripe
50 Albert Road
Richmond
Surrey TW10 6DP

© 2013 Tiger of the Stripe
All rights reserved

ISBN 978-1-904799-53-5

Typeset in the UK by
Tiger of the Stripe

Introduction

This book is, I believe, the first to include Richard Atkyns's *Vindication*, his *King's Grant for the Privilege for Sole Printing* and his *Original and Growth of Printing* in a single volume. The last-named is the best known of the three, celebrated more for its errors and sycophancy than for its historical value. The *Vindication*, although clearly self-serving in parts, is far more interesting, offering vivid descriptions of the battles in which Atkyns fought and insights into seventeenth-century gender politics.

In this edition the text of the *Original and Growth of Printing* is based on the 1664 quarto, British Library shelfmark 619.f.17 (1), rather than the earlier broadside, and that of the *Vindication* on the first edition, the 1669 quarto, British Library shelfmark 4902.cc.16. The *King's Grant* was also published in 1669 and the copy in St John's College, Cambridge, class mark Gg.3.20(10), provides the text for this edition. Editorial intervention has been kept to a minimum, with spelling and punctuation only changed when thought absolutely necessary. A few notes have been provided, largely identifying people and events which will not be familiar to most non-historians.

The Original and Growth of Printing

Richard Atkyns is largely remembered today as an early historian of printing. It is unfortunate that his *Original and Growth of Printing* is, at best, fanciful. This brief work, originally printed as a broadside and then in book form, had a clear purpose and one Atkyns did not seek to disguise: it was intended to prove that printing belonged to the sovereign in his public and private capacities 'as Supream Magistrate, and as Proprietor'. By showing this, Atkyns believed that he could establish that the royal patent he held for the printing of law textbooks was valid and inviolable.

Fanciful history of printing.

Royal patents.

INTRODUCTION

Atkyns was not satisfied with purely constitutional arguments to show the King's authority in this matter and he would have been foolish to do so. Royal patents, although a convenient way of augmenting the King's income, had long been a source of friction between the sovereign and parliament. Neither King nor Parliament was about to plunge England into another civil war over so trivial a matter. The reality of a constitutional monarchy had not escaped Atkyns's attention and he addressed his suit not only to the King but also to both Houses of Parliament. On the other hand, Atkyns had fought on the royalist side and he did not intend that Charles II should forgot it. In fact, Atkyns was an unreconstructed monarchist. He referred to Charles I's 'Most Just Cause,' appeared to support the moribund concept of the Divine Right of Kings, referring to Charles II's 'Sacred Person,' and suggested that any loss of royal power was 'a great Detriment to his Government'.

Unashamed sycophancy.

To the modern reader, the two striking characteristics of the *Original and Growth of Printing* are unashamed sycophancy and the lack of concern for historical accuracy. To be fair, neither was unusual at the time (and neither are they now). When a few men controlled the fortunes of many, toadying could be a useful, if not essential, tool.

The nub of Atkyns's argument was that the introduction of printing into England had been a royal project pre-dating Caxton's return from the Continent by a decade. This was a difficult story to support, especially since earlier historians, including the well-regarded John Stow, had already stated that printing had been brought to England by William Caxton in the 1470s. Atkyns thought it surprising that the new art had taken so long to reach England, although he supposed that this might be explained if, as Stow claimed, it was the act of a mercer (Caxton) rather than a more eminent person. In this, Atkyns displayed another of his less endearing characteristics, a relentless snobbery which elsewhere prompts him to describe fiddlers as 'despicable' simply because they play for a living.

Relentless snobbery.

Rufinus: Expositio Sancti Ieronimi.

INTRODUCTION

Atkyns was lucky enough to have come across a book (see p. 13) which seemed to disprove the date given for the introduction of printing into England by earlier historians – and if the date was wrong, perhaps, too, was the story of how printing came to England. A cynic might suspect that Atkyns had invented this book to suit his purpose but in fact it is genuine, although it is accepted today that it is misdated. Such misdating was surprisingly common in early books. It is believed that it was actually printed in 1478, not 1468.

While Atkyns had every incentive to accept the date at face value, it is hard to see that any impartial person would have come to a different conclusion at the time – it seemed to prove that printing had been brought to England by someone before Caxton's return to England in the 1470s. However, around this piece of genuine (if misleading) evidence, Atkyns then manufactured a rather implausible story in which Thomas Bourchier, Archbishop of Canterbury, persuaded Henry VI to support his plan for obtaining the secret art of printing. This was achieved by bribing one Frederick Corsellis, a workman, to desert his post at Gutenberg's printing works in Haarlem (where Atkyns says Gutenberg invented printing).² Corsellis was taken under guard to Oxford where he set up the first printing press outside Haarlem and Mainz. All this, according to Atkyns, is contained in a document held at Lambeth Palace which he hopes to publish at a later date.

William Caxton's return to England.

Corsellis bribed to come to England.

The involvement of Henry VI in this piece of industrial espionage was clearly intended to established the monarchy's right to regulate printing, an essential element in Atkyns's efforts to protect his monopoly of law textbooks from the claims of the Stationers' Company. No doubt his description of the role of the Archbishop of Canterbury was intended to curry favour with the clerical lobby.

As well as emphasising the King's absolute right to control printing, Atkyns was anxious to show that the Stationers' Company was totally unfit to be given so much power over the printing and publishing trades. The Company's rights, like Atkyns's, were conferred

Stationers' Company not fit to control printing.

INTRODUCTION

upon it by letters patent from the King. If Atkyns had been a less partial witness, he might have conceded that this proved the danger of the patent system and the royal prerogative in such matters. Driven largely by the desire of monarchs to raise money without Parliament's permission, letters patent had produced a confusing mess of overlapping rights which often brought riches to the legal profession and impoverished the book trade. It only weakened his case that he was asking Parliament (which disapproved of letters patent for their very extra-parliamentary nature) and the King to overthrow another set of royally-granted privileges for his own.

Atkyns argued that it was in the sovereign's interest to curtail the power of the Stationers. Before the Stationers' Company existed, he said, nothing could be printed without 'the Kings especiall Leave and Command,' but after its foundation, the trade simply published whatever was most profitable. The Company, far from suppressing abuses of the trade as it was intended to, allowed even greater abuses to take place and 'the Paper-pellets became as dangerous as Bullets.' In particular, the Company had done nothing to suppress the many books supporting the parliamentary side before and during the Civil War, including a number calling for and justifying the execution of Charles I.

Stationers' Company encourages abuse of printing.

The King's Grant of Privilege

Although arguing the same cause as the *Original and Growth,* this pamphlet is rather more subtle. Instead of snobbery and sycophancy, Atkyns presents his case as a legal one, larding his case with the *dicta* of his foresaken trade. It is not closely argued and is unlikely to have swayed legal minds but in matters of royal privilege Atkyns may have thought this hardly mattered.

The Vindication of Richard Atkyns

The *Vindication* is a short autobiography covering everything from Atkyns's ancestry, birth and education to his wooing of Martha

INTRODUCTION

Acheson, his part in the English Civil War, his battles with his creditors, the collapse of his marriage and the legal battle with the Stationers' Company. It may be as self-serving as the *Original and Growth of Printing* but it is not completely one-sided with regard to his marital relations. Although he portrays his wife's treatment of him in latter years as vicious and dishonest, he readily admits her great kindness to him in the early part of their marriage.

While he feels ill-treated by his wife, his creditors and others, Atkyns is not unaware of his own failings, including a youthful fondness for alcohol and a certain financial recklessness. Although not cited as an example of his extravagance, his loss of a £300 (about £41,700 today based on purchasing power)³ diamond-encrusted hatband (p. 67) is very telling. Nonetheless, an impartial observer would probably conclude that Atkyns, like so many others, would have remained solvent if he had fought on what was (at least, until the Restoration) the winning side. Without the financial strains placed upon it, his marriage might also have survived.

Atkyns sometimes seems to regard women as second-class citizens (for instance, quoting the opinion of a 'sober divine' that a man can divorce an adulterous wife without her consent, while the opposite case 'is not altogether so clear'), but in this he is only reflecting the mores and, indeed, the law of the time. A *feme sole* (single woman) in English law could maintain control of her own property (although it was often held in trust for her) but a *feme covert* (married woman) lost all property rights, the couple being a single legal entity with the husband controlling all property. Martha Atkyns was, arguably, a double victim of this inequitable system. That little property which she held in her own right after the death of her first husband automatically became her second husband's when she remarried, but most of the estate inherited from her first husband and her father was placed in trust for her. This was intended to protect her by preventing her husband from squandering her property but, if Atkyns is to believed, her uncle abused his position as trustee, granting himself a twenty-one year

Atkyns aware of his own failings.

Atkyns's attitude to women.

The position of married women in English law.

[ix]

INTRODUCTION

lease on her property in the Strand (p. 69). In doing so, he deprived both Atkyns and his wife of a valuable asset. Atkyns's frustration at not being able to do what he wished with what he considered to be his (or at least *their*) property may have coloured his judgement of her uncle and the other trustees. As he makes clear, it is undoubtedly true that a husband in those days who allowed his wife a substantial degree of financial freedom was risking everything, for any debts she ran up were his in law; and according to Atkyns, he was imprisoned for debts his wife had incurred.

By the standards of his time, Atkyns does not seem to have been particularly sexist and, while he rails against 'Government in Women Covert,' he does at least concede, '… where men are Beasts (which I hope is not in this Case) I have known several discreet Women that have preserved their families from ruine.'

Atkyns's accounts of the Civil War are not those of a professional soldier but he was often close to the centre of the action and (thanks, at least in part, to his family connections) intimate with a number of the senior officers. Even if Richard and Martha's treatment by parliamentary authorities and the ruining of their estates caused by billeting, neglect and litigation come as no surprise, his descriptions of their tribulations during and after the Civil War shed some light on the difficulties of everyday life during that period; every revolution has winners and losers and the Atkynses, like many other royalists, were losers, few recovering the wealth they had enjoyed before the war, even with the Restoration.

As might be expected, much of the *Vindication* is devoted to Atkyns's battles with the Stationers and with his creditors. He no doubt expected that the Restoration would allow him to re-assert his royal patent but he had reckoned without the permanent shift in power from the monarchy on the one hand and Parliament and the Stationers' Company on the other. This shift is typified by the ability of Henry Hills, former Printer to 'His Highness the Lord Protector,' to become Printer to Charles II and Warden of the Stationers' Company.[4] In such a climate, it was inevitable that Atkyns

INTRODUCTION

would struggle to wrest his monopoly for printing law textbooks from the Stationers and regain his fortune.

As well as his sometimes valuable and sometimes (in the case of the *Original and Growth*) worthless contributions to the historical record, Atkyns has bequeathed us one of the most tantalising (and least used) expressions in English literature, 'countesses of the better sort' (p. 67). Who, one wonders, were the worse sort of countesses?

'Tiger'
Richmond, December 2012

Contents

Introduction   v

The Original and Growth of Printing   v

The King's Grant of Privilege   viii

The Vindication of Richard Atkyns   viii

𝕿𝖍𝖊 𝕺𝖗𝖎𝖌𝖎𝖓𝖆𝖑 𝖆𝖓𝖉 𝕲𝖗𝖔𝖜𝖙𝖍 𝖔𝖋 𝕻𝖗𝖎𝖓𝖙𝖎𝖓𝖌   1

𝕿𝖍𝖊 𝕶𝖎𝖓𝖌'𝖘 𝕲𝖗𝖆𝖓𝖙 𝖔𝖋 𝕻𝖗𝖎𝖛𝖎𝖑𝖊𝖌𝖊   37

𝕿𝖍𝖊 𝖁𝖎𝖓𝖉𝖎𝖈𝖆𝖙𝖎𝖔𝖓 𝖔𝖋 𝕽𝖎𝖈𝖍𝖆𝖗𝖉 𝕬𝖙𝖐𝖞𝖓𝖘 𝕰𝖘𝖖𝖚𝖎𝖗𝖊   51

Notes   129

Bibliography   140

Index   143

THE
Original and Growth
OF
PRINTING

Collected
Out of *HISTORY*, and the *Records*
of this *KINGDOME*.

Wherein is also Demonstrated,

That PRINTING appertaineth to the
Prerogative *Royal;* and is a Flower
of the *Crown* of *England*.

By RICHARD ATKYNS, *Esq.*

𝕎𝕙𝕚𝕥𝕖-𝕙𝕒𝕝𝕝. *April* the 25[th.] 1664
By 𝕺𝖗𝖉𝖊𝖗 and 𝕬𝖕𝖕𝖔𝖎𝖓𝖙𝖒𝖊𝖓𝖙 of the Right HONORABLE,
Mr. 𝕾𝖊𝖈𝖗𝖊𝖙𝖆𝖗𝖞 Morice, Let this be 𝕻𝖗𝖎𝖓𝖙𝖊𝖉.

THO: RYCHAUT.

RICHMOND:
Printed for TIGER OF THE STRIPE,
MMXIII

TO THE KINGS MOST Excellent Maiesty.

Most Gracious and Dread Soveraign,

Though I had the Honour to be very well known to His Majesty of ever Blessed Memory, Your most Royall Father, and to be a Sufferer in the loss of a considerable Estate, for His Most Just Cause, yet I may not be so well known to Your Sacred Person: however, the same Duty that moved Me to fight for Him, remains in Me to write for You; not out of any Confidence in my Pen (for I am the first shall judge that my Self) but out of Conscience and Loyalty to my Soveraigne; for whose sake, I resolve to hazard Censure, rather than to be wanting in any Discovery, that may tend to your Majesties Interest, and indubitate Right.

The least loss of Power in a Magistrate, is a great Detriment to his Government, and an Advantage to his Enemies; the least Creep-Window robs the whole House; the least Errour in War not to be redeem'd: And as that ever Blessed late Martyr said (when He gave his Watch of Government, to be cleansed by the too-long Parliament) the least Pin of it being left out, would cause a Discord in the whole: Therefore might Solomon well say, Where the Word of a King is there is Power: *The* King *and* Power being Relatives.

RICHARD ATKYNS ESQUIRE

That Printing *belongs to Your Majesty, in Your publique and private Capacity, as Supream Magistrate, and as Proprietor, I do with all boldness affirm; and that it is a considerable* Branch *of the* Regal Power, *will no Person deny: for it ties, and unties the very Hearts of the People, as please the Author: If the* Tongue, *that is but a little Member, can set the Course of Nature on Fire; how much more the* Quill, *which is of a flying Nature in it self; and so Spiritual, that it is in all Places at the same time; and so Powerful, when it is cunningly handled, that it is the Peoples Deity.*

That this Power which is intire and inherent in Your Majesties Person, and inseperable from Your Crown, should be divided, and divolve upon Your Officers (though never so great and Good) may be of dangerous Consequence: You are the Head of the Church, and Supream of the Law; shall the Body Govern the Head? Men use to trust, when they cannot avoid it; but that there may be a Derivative and Ministerial Power in them, with Appeal to Your Majesty, I do with all Humility admit and propose.

Printing *is like a good Dish of Meat, which moderately eaten of, turns to the Nourishment and health of the Body; but immoderately, to Surfeits and Sicknesses: As the Use is very necessary, the Abuse is very dangerous: Cannot this Abuse be remedied any other way, then by depriving Your Majesty of Your Antient and Just Power? How were the Abuses taken away in Queen* Elizabeth, *King* James, *and the beginning of King* Charles *his time, when few or no Scandals or Libels were stirring? Was it not by Fining, Imprisoning, Seizing the Books, and breaking the* Presses *of Transgressors, by Order of* Councel-Board?[5] *Was it not otherwise when the Jurisdiction of that Court was taken away by* Act of Parliament, 17 Car. *If Princes cannot redress Abuses, can less Men redress them? I dare positively say, the* Liberty *of the Press, was the principal furthering Cause of the Confinement of Your most Royal Fathers Person; for, after this* Act, *every* Male-content *vented his Passion in Print; Some against his Person, some against his Government, some against his Religion, and some against his Parts: the Common People that before this Liberty believed even a* Ballad, *because it was in Print, greedily suckt in these Scandals, especially being Authorized by a God of their own making: the* Parliament *finding the Faith of the* Deceived People *to be implicitly in them,* Printed the Remonstrance,⁶ the Engagement to live and dye with the Earl of Essex,⁷ the Covenant,⁸ &c. *and so totally possest the Press that the King could not be heard: By this means the* Common People *became not onely* Statists, *but* Parties *in the* Parliaments *Cause, hearing but one side, and then* Words *begat* Blows: *for though Words of themselves are too weak Instruments to Kill a Man; yet they can direct how, and when, and what Men shall be killed: In the Statute of* 21 Jac. *printing keeps very able Company, as* Salt-Peter, Gun-Powder, Ordnance, &c. *all which are Exempted from being* Monopolies.⁹

Not to be longer tedious, I too much fear, this late Act for two years compleats all the former Concessions of the late King: I know it was done in hast, and with a good Intent; but by Your Majesties *Gracious Leave and Pardon, even then very considerable Persons in Your* House of Commons, *were of Opinion they had nothing to do with it, the Power of the Press being so wholly in Your* Majesty. *Indeed, Necessity that hath no Law, was the cause of this Law,* viz. *to hinder the Growth of Scandalous* Books *and* Pamphlets; *but it hath fallen very short of the End: for few or none, of many Printed, have bin bought in by the* Stationers. *I have now discharged my Duty to Your* Majesty, *and if I find I have so far prevailed upon Your Royall Goodness, as to ask unconcern'd Councel what is best to be done, I have my End; I hope Your* Majesty *will have the Advantage. So prayeth,*

                Your Sacred Majesties

                    most Humble Servant,

                        and most obedient Subject,

                          RICHARD ATKYNS.

TO
The Right Honourable,
THE
LORDS
AND TO
The Honourable,
THE
COMMONS
ASSEMBLED IN
PARLIAMENT.

May it please your Honours,

I *Have ever better underſtood mine own Disabilities, than to desire to appear in Print, where the Author ſtands as a* Butt *to be shot at, by the sharp Arrows of every busie* Critick, *and runs a moſt certain hazard, and moſt uncertain Benefit: But having been above twenty three years in* Chancery, *and other Courts of Juſtice; and ſpent more then One Thousand Pounds, in vindicating the Kings Grant of Printing the Common Law of* England, *and his Lawful Power to grant the same, and kept His*

RICHARD ATKYNS ESQUIRE

Title alive even in the worst of Times (when 'twas reputed unlawful, because the Kings) I cannot refrain from defending it, now the King is, or ought to be, restored to His Rights again; especially since all Persons are invited by Order to speak their Minds freely concerning the Subject: So that there is a Necessity upon me to speak now, or for ever hereafter to hold my Peace; this being probably the last time of Asking.

'Tis not unknown to every Member of each House, how little benefit hath accrued to the Kingdom, by the late Act of Parliament for two years, Entituled, An A C T for preventing the frequent Abuses in Printing Seditious, Treasonable, and unlicensed Books and Pamphlets, &c. Which Act determines June next: Nor can it be thought, but that there is Cause enough for another Act to take place, when this is expired. The Reason why this present Act hath operated so little is most apparent; because the Executive Power is plac'd in the Company of Stationers, *who onely can offend, and whose Interest is to do so: They are both Parties and Judges, and 'twere a high Point of Self-denial for Men to punish themselves: But they will wipe their Mouthes with* Solomon's *Harlot, and take it very unkindly, if the same, or a greater Power be not continued to them in the next Act to be made: They will promise as fair as the Long Parliament did to the late King* (to make Him a Glorious King) *and perform it as certainly as they did too.*

Jugglers seldome shew the same Trick twice together; and the Italian *Proverb is,* If a man deceives me once, 'tis his fault; if twice, it is mine own: *That the Great Councel of this Nation, should further trust those that have deceived them already, and believe fair Pretences, contrary to Reason and Practice, would be a sad Fate upon us all; when wofull Experience tells us, That if the King be taken from being Head of the Law, there will not want a Law to take off His Head in a short time.*

There were a sort of People in King David's Time, which imagined Mischief as a Law; as in the late King's time, that practised Mischief by a Law: which might incline the Parliament to frame a strict Law against this kind of Mischief. But I hope the King's Mercy in forgiving such, (by which He imitates His Maker) will find so hearty a Conversion, that Ingratitude shall never joyn with Rebellion, to provoke a Tyrranical Government over this Kingdom: such Men (if I may so call them) are worse then the Gentiles, *of whom St.* Paul *saith,* That having not the Law, and doing by Nature the things contained in the Law, are a Law unto themselves, which shew the Law written in their Hearts; *nay, worse then Beasts, who by Nature observe a Law amongst themselves.*

Shall sense and Reason alone teach Creatures willingly to confine themselves to certain Rules for the Common Good, and shall Professors of Christianity break them? Shall the Law of Nature command Men to be free from offending; and shall the Law of God be thought to command them to be free to offend? Let not our too-near Neighbours the Turks *have that Advantage against us. But whilest I declaim against others for breaking their Bounds, I may be thought guilty of committing the same Errour myself; I shall therefore most humbly beg your Honours Pardon, and rest,*

Your Honours

Most Humble, and Faithful

Servant,

RICHARD ATKYNS.

THE
Originall and Grovvth
OF
PRINTING.

REASON is the great Distinction between Man and Beast; *Gusman* calls the Man of most Knowledg, *A God amongst Men*. And Bishop *Hall*[10] divides the whole Duty of Man into *Knowledg* and *Practice*. In the Infancy of the World (especially before the Sealing of the Scripture-Canon) God Revealed himself and his Will frequently, either Vocally by himself, as to *Moses* in the Mount; or else by divers and Sundry other manners, As by Dreams, Visions, Prophecies, Extasies, Oracles, and other Supernatural means: Nor will I Blow up the Humours of these Times so high, as to Confine these his Miraculous Revelations to Gods People onely (though to them most frequently and especially), but sometimes also to Hypocrites within the Church, as to *Saul* and others; yea and sometimes even to Infidels, as to *Pharaoh, Balaam, Nebuchadnezzar, Abimelech*, &c. But since the Writings of the Prophets and Apostles, (commonly called the Scriptures) And that the Christian Church by the Preaching of the Gospel, is become Oecumenical, Dreams and other Supernaturall Revelations, as also other things of like nature as Miracles, have ceased to be of ordinary and familiar use; So as now we ought rather to suspect Delusion in them, than to expect Direction from them: Yet God hath no where abridged or Limited himself from these supernatural wayes of Revealing his

Revelation not Confined only to the People of God.

RICHARD ATKYNS ESQUIRE

Will, in case his Written Word Should be taken from us, or we from it: But we of this Latter Age have all these so Lively represented to our View, by the benefit of *Printing,* as if we our selves were personally present: For *Printing* is of so Divine a Nature, that it makes a Thousand years but as yesterday, by Presenting to our View things done so long before; and so Spirituall withall, that it flyes into all parts parts of the World without Weariness. Finally, 'tis so great a Friend to the Schollar, that he may make himself Master of any Art or Science that hath been treated of for 2000 years before, in lesse than two years time. But Virtue it self will not want Opposers, and Philosophy is ever odious to ignorant Ears: Nay, there are a sort or People in the World, that account Ignorance the Mother of Devotion, and therefore out of Conscience would not have even the Scriptures Printed in the Mother Tongue:[11] But I shall not go out of my own Way, to bring them unto it, further than by defending the Theame I have in hand.

The great Benefit of Printing.

Concerning the time of bringing this Excellent A R T into England, and by whose Expence and Procurement it was brought; Modern Writers of good Reputation do most erroniously agree together. Mr. *Stowe* in his *Survey of London*[12] speaking of the 37th year of King *Henry* the Sixth his Reign, which was Anno Dom. 1459. saith, That the Noble Science of P R I N T I N G was about this time found in Germany at *Magunce* by one *John Cuthenbergus*[13] a Knight, And that William Caxton of London, Mercer, brought it into England about the Year 1471. And first practised the same in the Abby of St. *Peter* at *Westminster;* With whom Sir *Richard Baker* in his Chronicle[14] agrees throughout. And Mr. *Howell* in his *Historical Discourse of London* and *Westminster*[15] agrees with both the former in the Time, Person, and Place in general; but more particularly declares the Place in *Westminster* to be the Almonry there; And that *Islip*[16] Abbot of *Westminster* set up the first Press of Book-Printing that ever was in *England*. These three famous Historians having fill'd the World with the supposed truth of this Asser-

Printing supposed to be brought into England in the Year 1471.

Page 404.
Page 284.

Page 353.

[12]

tion, (Although possibly it might arise through the mistake of the first Writer only, whose Memory I perfectly honour) makes it the harder Task upon me to undeceive the World again: Nor would I undertake this Work, but under a double notion: As I am a Friend to Truth, and so it is unfit to suffer one Man to be intituled to the worthy Atchievements of another. And as a Friend to my Self, not to lose one of my best Arguments of Intituling the King to this A R T in his Private Capacity.

Historians must of necessity take many things upon trust, they cannot with their own but with the Eyes of others see what things were done before they themselves were, *Bernardus non vidit omnia;* 'Tis not then impossible they should mistake. I shall now make it appear they have done so, from their Own, as well as from other Arguments: Mr. *Stowe* his Expressions are very dubious, and the matter exprest very Improbable; He saith *P R I N T I N G* was found in *Magunce,* which presupposes it was practised some where else before, and lost: And further, That 'twas found in the Reign of *Henry* the Sixth, *Anno Dom.* 1459. and not brought into *England* till Eleven years in the succeeding Reign of *Edward* the Fourth, being 12 years after, as if it had been lost again. If this be true, there was as little Rarity as Expedition in obtaining it, the age of 12 years time having intervened, and so indeed it might be the Act of a Mercer rather than a more eminent Person: But when I consider what great advantage the Kingdome in general receives by it, I could not but think a Publique Person and a Publique Purse must needs be concerned in so publique a Good. The more I Considered of this, the more inquisitive I was to find out the truth of it: At last, a Book came to my hands Printed at *Oxon, Anno Dom.* 1468.[17] which was three years before any of the recited Authours would allow it to be in *England;* which gave me some reward for my Curiosity, and encouragement to proceed further: And in prosecution of this Discovery, the same most worthy Person who trusted me with the aforesaid Book, did also present me with the Copy of a

RICHARD ATKYNS ESQUIRE

Record and Manuscript in *Lambeth*-House,[18] heretofore in his Custody, belonging to the See (and not to any particular Arch-Bishop of *Canterbury*); the Substance whereof was this, (though I hope, for publique satisfaction, the Record it self, in its due time, will appear.)

Thomas Bourchier, Arch-Bishop of *Canterbury* moved the then King (*Hen.* the 6th) to use all possible means for procuring a Printing-Mold (for so 'twas there called) to be brought into this Kingdom; the King (a good Man, and much given to Works of this Nature) readily hearkned to the Motion; and taking private Advice, how to effect His Design, concluded it could not be brought about without great Secrecy, and a considerable Sum of Money given to Such Person or Persons, as would draw off some of the Work-men from *Harlein* in *Holland,* where *John Cuthenberg* had newly invented it,[19] and was himself personally at Work: 'Twas resolv'd, that less then one Thousand Marks would not produce the desir'd Effect: Towards which Sum, the said Arch-Bishop presented the King with Three Hundred Marks. The Money being now prepared, the Management of the Design was committed to Mr. *Robert Turnour,*[20] who then was of the Roabs to the King, and a Person most in Favour with Him, of any of his Condition: Mr. *Turnour* took to his Assistance Mr. *Caxton,* a Citizen of good Abilities, who Trading much into *Holland,* might be a Creditable Pretence, as well for his going, as stay in the Low-Countries: Mr. *Turnour* was in Disguise (his Beard and Hair shaven quite off) but Mr. *Caxton* appeared known and publique. They having received the said Sum of One Thousand Marks, went first to *Amsterdam,* then to *Leyden,* not daring to enter *Harlein* it self; for the Town was very jealous, having imprisoned and apprehended divers Persons, who came from other Parts for the same purpose: They staid till they had spent the whole One Thousand Marks in Gifts and Expences: So as the King was fain to send Five Hundred Marks more, Mr. *Turnour* having written to the King, that he had almost done his Work; a Bargain (as he said) being struck betwixt him and two *Hollanders,* for bringing off one of the Work-men, who should sufficiently discover and teach this New Art: At last, with much ado, they got off one of the Under-Workmen, whose Name was Frederick *Corsells* (or rather *Corsellis*),[21] who late one Night stole from his Fellows in Disguise, into a Vessel prepared before for that purpose; and so the Wind (favouring the Design) brought him safe to *London.*

'Twas not thought so prudent, to set him on Work at *London,* (but by the Arch-Bishops meanes, who had been Vice-Chancellor, and afterwards Chancellor of the University of *Oxon*) Corsellis was carryed with a Guard to *Oxon;* which Guard constantly watch'd, to prevent *Corsellis* from any possible Escape, till he had made good his Promise, in teaching how to Print: So that at *Oxford* Printing was first set up in *England,* which was before there was any Printing-Press, or Printer, in *France, Spain, Italy, or Germany,* (except the City of *Mentz*) which claimes Seniority, as to Printing, even of *Harlein* it self, calling her City, Urbem Maguntinam Artis Tipographicæ Inventricem primam, though 'tis known to be otherwise, that City gaining that Art by the Brother of one of the Workmen of *Harlein,* who had learnt it at Home of his Brother, and after set up for himself at *Mentz.* *Printing first set up at Oxford.*

This Press at *Oxon* was at least ten years before there was any Printing in *Europe* (except at *Harlein,* and *Mentz*) where also it was but new born. This Press at *Oxford,* was afterwards found inconvenient, to be the sole Printing-place of *England,* as being too far from *London,* and the Sea: Whereupon the King set up a Press at St. *Albans,*[22] and another in the Abby of *Westminster,* where they Printed Several Bookes of Divinity and Physick, (for the King, for Reasons best known to himself and Council) permitted then no Law-Books to be Printed; nor did any Printer exercise that A R T, but onely such as were the Kings sworn Servants; the King himself having the Price and Emolument for Printing Books. *Printing depraved by being Incorporated with others.*

RICHARD ATKYNS ESQUIRE

None but the Kings sworn servants permitted to be Printers.

Printing thus brought into *England,* was most Graciously received by the King, and most cordially entertained by the Church, the Printers having the Honour to be Sworn the King's Servants, and the Favour to Lodge in the very Bosome of the Church; as in *Westminster,* St. *Albans, Oxon,* &c. By this meanes the A R T grew so famous, that *Anno prim. Rich. 3, cap. 9.* when an Act of *Parliament* was made for Restraint of Aliens, from using any Handicrafts here (except as Servants to Natives) a special *Provisoe* was *inserted,* that Strangers might bring in Printed or Written Books, to sell at their pleasure, and Exercise the A R T of Printing here, notwithstanding that Act: So that in the space of 40 or 50 years, by the especial Industry and Indulgence of *Edw.* the Fourth, *Edw.* the Fifth, *Rich.* the Third, *Henry* the Seventh, and *Henry* the Eighth, the *English* prov'd so good Proficients in Printing, and grew so numerous, as to furnish the Kingdome with Books; and so Skilfull, as to print them as well as any beyond the Seas, as appears by the Act of the 25 *Hen.* 8. *cap.* 15. which Abrogates the Said *Provisoe* for that Reason. And it was further Enacted in the said Statute, That if any person bought Forreign Books bound, he should pay 6 *s.* 8 *d. per* Book. And it was further Provided and Enacted, That in case the said Printers and Sellers of Books, were unreasonable in their prices, they should be moderated by the Lord Chancellor, Lord Treasurer, the two Lord Chief Justices,[23] or any two of them, who also had power to Fine them 3 *s.* 4 *d.* for every Book whose price shall be enhanced.

The Price of Books no to be enhanced.

Thus was the A R T of Printing, in its Infancy, Nursed up by the Nursing Father of us all, and in its riper Age brought up in Monasteries of greatest Accompt; and yet were the Instruments thereof restrained from the Evil of enhancing the prices of Books, to the Detriment of their Fellow-Subjects, by the Authority aforesaid. While they had this Check upon them, they were not only Servants to the King, but Friends to the Kingdom; But when they were by Charter Concorporated with *Book-Binders, Book-Sellers,* and *Found-*

ers of Letters, 3 and. 4 *Phil.* and *Mary*, and called the Company of Stationers,[24] the Body forgot the Head, and by degrees, (breaking the Reines of Government) they kickt against the Power that gave them Life: And whereas before they Printed nothing but by the Kings especiall Leave and Command, they now (being free) set up for themselves to print what they could get most Money by; and taking the Advantage of those Virtiginous Times, of the latter end of *Henry* the 8. *Edward* the 6. and Queen *Mary*, they fill'd the Kingdom with so many Books, and the Brains of the People with so many contrary Opinions, that these Paper-pellets became as dangerous as Bullets, to verifie that Saying of *Tertullion, That Lawyers Gowns hurt the Common-wealth as much as Souldiers Helmets.* Thus was this excellent and desireable A R T, within less than one hundred years, so totally vitiated, that whereas they were before the King's Printers and Servants, they now grew so poor, so numerous, and contemptible, by being Concorporated, that they turn'd this famous A R T into a Mechanick Trade for a Livelyhood.

But here I must break off (though abruptly) and answer an Objection; for methinks I hear the *Critick* say, *How can that be a Mechanick Trade now, that the Author allowes to be a famous Art heretofore, being alwayes one and the same thing?* *Object.* 1.

The Matter of which before I answer, I must crave leave to give you the signification of the Word *Mechanick;* the rather, because the several sorts of Trades, of which the Company of *Stationers* are Composed (and more particularly the *Book-Sellers* who say they are of no Manufacture) do peremptorily deny themselves to be Mechanicks.

The Word *Mechanicus*, which signifies a Handicrafts-man, doth in the strict sense comprehend *Printers, Founders of Letters* and *Book-Binders;* And I believe, in the large Sense, all Trades-men whatso-

How and why the ART is called a Mechanick Trade.

[17]

Answ. 1.

ever: But if that be deficient, let us go to the Original Greek word μηχανὰ, which signifies a *Cunning Contrivance* of the Head, as well as Hand; and this will certainly take in all Trades, for as much as there is *Cunning* in all Trades: But if it should miss any, yet it cannot fail of the *Company of Stationers,* because they are denominated a *Mystery,*[25] and there the strict signification of the Word comes in again.

Now for the matter of the Objection, *That a famous A R T cannot be a Mechanick Trade.*

Answ. 2.

I Answer, This is so far from being true, that there is nothing in Nature but is good or bad according as 'tis us'd; for the great Creator of all things made nothing to no purpose; even Meat and Drink (without which we cannot live) if abus'd, destroyes life: Twenty dye of Surfets, for one that is starved for want of Meat. But to give you an instance *ad idem:*

A simile taken from Musick.

Musick is not onely an Art, but one of the Liberall Arts practised by Princes themselves, and made instrumentall to the Glory of God; yet what Trade is there more despicable in the World both in Name and Nature, than a Common Fidler; though he may draw as good a sound out of an Instrument, and have as much Art in Playing and Composing as any Gentleman; yet if he get his Living by it, and makes it his Trade, he is still but a Fidler: and herein I pity him more than any of other Professions, because he perverts the Creation, and turns Day into Night; for most commonly when sober Persons are in Bed, he must play to please the humours of the lighter sort; And though his Heart be ready to break through Melancholy, he must sing a merry Song to delight the Company, if commanded, or have his Fiddle sing about, his Ears: Is not this Mechanick, think you?

But to Return where I digrest; *Printing* became now so dangerous to the Common-wealth, That there were more Books Burnt in Ten years, than could be Printed in Twenty: And now it concern'd the

THE ORIGINAL AND GROWTH OF PRINTING

Prince altogether as much to Suppress the Abuse, as it was before to Obtain the Use of Printing; And had there not been a Reserve of Licensing Such Books as should be Printed still remaining in the Crown, they might have published the wickedness of their own Imaginations with Authority. But Queen *Elizabeth* at her very first Entrance to the Crown, finding so great Disorders in *Church* and *State,* by reason of the abuse in *Printing* Secures in the first place the *Law* and the *Gospel,* of both which the Kings and Queens of *England* have inherent Right as Heads of the Church, and Supream of the Law; and not onely in their publique, but private Capacity, as Proprietors; the Power and Signiory of this, under Favour, cannot be severed from the Crown: The Kings being the Trustees of the People, who have formerly taken an Oath at their Coronation, *That they shall keep all the Lands, Honours, and Dignities, Rights, and freedoms of the Crown of* England, *in all manner whole, without any manner of minishment; and the Right of the Crown, hurt, decay'd or lost to their Power shall call again into the Antient Estate.* Which Oath, the said Queen kept inviolably, and liv'd the more quietly for it all the time of her Reign, and died in Peace. True it is, they may, and do gratifie their Friends and Servants, in giving them the Emoluments and Profits that arise from *Printing;* but the Power they cannot alienate from the Crown, without losing the most pretious Stone out of their Diadem. To shew you one Example for all, the said Queen, the first Year of her Reign, grants by Patent the Priviledge of sole *Printing* all Books, that touch or concern the Common-Laws of *England,* to *Tottel*[26] a Servant to her Majesty, who kept it intire to his Death: After him, to one *Yestweirt,*[27] another Servant to Her Majesty: After him, to *Weight* and *Norton;*[28] and after them, King *James* grants the same Priviledge to *More,*[29] one of His Majesties Clerks of the Signet; which Grant continues to this Day; and so for the Bible, the Statute-Laws, the Book of Common-Prayer, Proclamations, as much as the Grammer, the Primer. *&c.* all granted by Kings and Queens, not onely to gratifie their Friends and

Patents for Printing, granted to several persons.

Servants, but to preserve the Regal Power and Authority on Foot, and these Books from being corrupted.

The Truth of this the most impudent Opponent will not deny, because the *Patents* themselves give Evidence against them; nor will they deny in words (though they do daily in fact) that the King hath Power to make such Grants.

Object. 2   But this they will Object and say, *That Gentlemen being not Printers by Trade, nor Free of the Company of Stationers, can never find out the Abuses of* Printing *themselves, nor understand the Cheats of them, they being so many; but they must be discovered either by the* Printers, *or the Company of* Stationers, *or both together:* This is the Common Objection.

Answer.   To which I Answer; The Objection cannot properly lye against any man for being a Gentleman, because the greatest Nobleman will not deny himself to be one, nor can he with Honour refuse a *Challenge* from any Gentleman: And the very Mechanick is so willing to disguise his want of Gentility, that when he arrives to a Considerable Estate, he is very forward to purchase Honour. Nor can I think any man the less knowing for being a Gentleman, whose Education is most commonly at School, at the University, the Inns of Court, Travell, or both; Whereas the Education of a Mechanick is only at School, without any other Improvements: This being the course that each of them generally runs, 'tis strange if the Gentleman should not get the start, and be better known to Letters, Manners, and Men, than the Mechanick, But this *Objection* goes further.

3. *Object.*   *That, though they may know Letters, &c. better than, the Mechanick, yet they can never arrive to a full Discovery of the Mystery, and deceitfull part of the Trade; that they must give Handicrafts men leave to know best.*

1. *Answ.*   To which I Answer; First, That there is no Magick in this Art; Jugglers they may be, but Conjurers they are not.

Secondly, That Gentlemen may and do know the Mystery and Deceipt of the Trade as well as those that act it; but their knowledg tends different wayes. It is the Gentleman or Patentees part to detect and hinder this Deceipt; As 'tis the *Stationers* to promote and practice it: Their Profit blinds them so, that they resemble certain Birds, who when they hide their heads, think none can see their Bodies;[30] Or like Children, who after a fault Committed, wink themselves, thinking thereby that none can find them out. To render this possible, I will give you an instance of a Person, that none can deny to be a Gentleman, though he were much more, (I mean the late King) who was not onely *aliquis in omnibus*, but *singularis in omnibus*.[31]

2. *Answ.*

This excellent Prince, hearing of a rare Head, amongst several other Pictures, brought *Me* from *Rome*, sent Sir *James Palmer*,[32] to bring it to *Whitehall* to Him, where were present divers *Picture-Drawers* and *Painters:* He ask'd them all, *Of whose Hand that was?* Some guest at it; Others were of another Opinion; but none was positive: At last, said the King, *This is such a Man's Hand, I know it as well, as if I had seen him draw it: But* (said he) *is there but one Man's Hand in this Picture?*

The King more skilfull than Mechanicks in their own Trade.

None could discern, whether there was or not: But most concluded, there was but one Hand: Said He, *I am sure there are two Hands in it, for I know the Hand that drew the Heads; but the Hand that drew the rest, I never saw before.* Upon this, a Gentleman that had been at *Rome*, about Ten Years before, affirmed, *That he saw this very Picture, with the two Heads, unfinished at that time; And that he heard his Brother* (who staid there some years after him) *say, that the Widow of the Painter that drew it wanting Money, got the best Master she could find to finish it, and make it saleable.* Is it not strange, that the King that was no *Picture-Drawer* himself, should see further into a Picture, than Painters by Trade.

RICHARD ATKYNS ESQUIRE

Patentees fittest to redress the Evils of the Press.

But were the Objection true (as 'tis much to be doubted) yet were the *Patentees* still the very fittest Persons to be imployed, in redressing the Evils of the Press, wherein they are concern'd, because their Interest leads them to it: And Men will come to a soon Discovery, even of obscure things, where their Interest inclines them; Indeed, the *Printers* Argument against the *Booksellers* &c. being all of the Company of *Stationers,* doth hold in point of Government amongst themselves, *That 'tis absurd and ridiculous for any, to have the Rule and Oversight of that which they have no insight in.*

A brief Discourse concerning Printing. Page 8.

But this is not at all applicable (nor do they intend it to be so) to the King's *Patentees;* who (if they be not *Printers* themselves, nor have a *Printer* of their own) agree with one to Print Such a Book, whereof they have the Propriety, which *Printer* gives him Security to Print the same perfect, and with a fair Letter; it matters not whether the *Pattentee* can set the Letters, or have Skill in the Manufacture himself; 'tis sufficient for him to examine it with his Copy when 'tis done, (which Copy cannot erre, because it is under the publique Licence) and try whether it be as 'twas agreed; and if it be not as it ought to be in all respects, the *Printer* loseth his Labour and Charge: 'Tis the Printers Interest then as well as the Patentees, to Print it perfect and fair; without which, (should they both joyn together) they could not vend it, after 'twas Printed. I confess, it would argue an ill Nature in me, not to be sorry for the just Occasion the *Printers* have to complain of their Brethren the *Booksellers,* were it not for this, *That when some men fall out, others shall hear of their Goods:* Yet I cannot but side with the *Printers* thus far, as to Declare, That they, with the Founders of Letters, are the onely Instruments of absolute Necessity in this A R T; whereas *Book-Sellers* might be Supply'd out of the She-Shopkeepers in *Westminster-Hall,* if all the rest were higher promoted. *In fine,* These *Book-Sellers* are the *Drones* that devour the Honey, made by the Laborious *Printers;* I cannot so sensibly express it, as themselves have done: therefore hear them, and not Me; Say they, *So far were the*

Printers and Founders onely necessary to the Art of Printing.

[22]

Stationers *from redressing the Printers Wrongs, that some of themselves, took upon themselves, the Exercise of their Function, and gave a Forreigner his Freedom* gratis, *to inable him to usurp the Exercise of the* Printers *Calling; and to compleat the Abuse, Erect a* Printing-House *of their own: so as it is become a Question among the* Book-Sellers, *Whether a* Printer *ought to have any Copy or no? Or if he have, They (keeping the Register) will hardly enter it: Or if they do, they and their Accomplices will use all means to disparage it, if not down-right counterfeit it, that they Tyrannize over* Printers; *And further, That for want of a due Establishment, Transgressors never want Incouragers to begin, or Chapmen to vend such Ware, when finished among the* Stationers. *They desire, that such as are free of the Trade, may be free indeed, and not manumitted (as of late) from the Service of one Master, to the Slavery of many Tyrants; That the* Stationers *have Usurped their Callings, and incouraged, yea hired others so to do, and stand related to each other, as the Buyer to the Seller. Upon all which, they refer their Cause, to the same* Power *that gave them theirs, who may resume, or abridge the same, upon Mis-use, at their pleasure.* This is a sad Complaint of Elder Brethren against their Younger; if one Dog will not prey upon another, what Reason can be given, why Men should devour Men? And if this be the Usage those must trust to, to Whom they profess Friendship; what is like to become of the *Patentees,* against whom they profess Emnity? If such a Power be continued to them, which I hope will be seriously considered of, before it shall be re-granted. Success (which usually gives Confidence) hath so hardned them, that having not felt the Justice of the King's Hand, for above twenty years last past, they now begin to swear Him out of, and Themselvs into, this Part of His Regal Power: For they being lately Examined upon Interrogatories between *Atkins et Uxor,* Plaintiffs, and *Flesher* and the *Stationers* Defendants; some of them (I am sure) are so streight-mouth'd, that they do not declare the whole Truth of what they know on our Part, and seem to make a Conscience of Swearing at all; As if St. *Paul* had been in an Errour, when he said, *An Oath for Confrmation, is an end of Strife.* Indeed, they strein'd at a Gnat, but

A Brief Discourse Concerning Printing Page 5.

Pag. 7.

Pag. 12.

Pag. 14.

The dangerous Consequence of Power in the Stationers.

The Stationers *Conscience.*

RICHARD ATKYNS ESQUIRE

when they were to Swear on the other Part (namely their own) they open their Mouthes wide enough to Swallow a Camel. *Say they, from the Year* 1641. *or* 1642. *until the time of His Majesties Blessed Restauration to His Crown, any* Booksellers *that listed, did print, or cause to be printed, such Law-Books as seemed good unto them, without Restraint or Prohibition, occasioned by the Licence of the late Times: And that such as had Licence under the Kings Grant to print Law-Books, were hindered to make the Benefit of the said Grant; And that it was usual for such persons as printed Law-Books, to enter the same in the Book of* Stationers-Hall; *And that it was conceived and taken, that such person and persons, as Entred a Copy in the said* Hall-Book *to be Printed had the sole Right to print the same; and those that claim'd the Right of Printing Law-Books under the Kings Licence, were thereby taken to be Excluded, and debarred to claim any Benefit therein.*

Observe what a Sad time the Kings Patentees endur'd for almost Twenty years together, confest by the Oaths of these honest Men, that joyn'd in dividing the Spoyl: And shall it be so still, now the King is returned again? I dare positively say it shall; Witness a Book called *Poultons Abridgment*,[33] particularly Exprest in the *Law-Patent,* which they Printed since the Kings Restauration, by vertue of an Entry only in their *Hall-Book,* against the said Patent; the Patentee's Lessee *Flesher*[34] (a principal Member of the Company) finding the *Stationers* like to be worsted at the Council-Board in the Contest of their said *Entry* against the Kings *Grant,* joyns Interest with them, and also Engageth the Kings Patentees Trustee, and the Kings Printers (who pretended Some Interest in the Said Book) on their side; As if severall bad Titles could create one good one: By which means after Four Hearings, the Cause (seeming to be between Party and Party) was dismist, but with the Recommendation of the King and Councill to the Lord High Chancellor, on the behalf of the Said *Patentee* (who had the Equitable Right): The Company to requite the Lessees kindness in defending them from a Contempt against the K I N G, (he being the person in

THE ORIGINAL AND GROWTH OF PRINTING

Law that fought to have defended the *Patent* for the King against the said company) like Brethren—joyn with the Lessee to defend him against the Justice of the Court of *Chancery*, and Combine together to defeat the Patentee of his Rent by Covenant, and so bandy the Legall Interest from one hand to another, that it cannot be yet found where it Vests; and to enable him the better, make him Master of the said Company for two years together (never known before), and choose Wardens fit for the purpose, who Engage the Stock of the whole Company on his behalf, against the King's Patentee: And being thus fortified, they published the said Book with this Title Page, *Printed for the Company of Stationers,* John Bill *and* Christopher Barker *his Majesties Printers;*[35] and so make a mixt Interest, to render the Title the more questionable in the future; but do not so much as mention the Kings Patent at all, by which the said Book is granted by Name: This serves the Turn for the present occasion, and being so possest as aforesaid, *Flesher* and the *Stationers* give 200 l. to the Patentee's Trustee to release the Rent and Covenants of the said free of Lease, and the Kings' Printers 100 l. or 200 l. for their assistance in so difficult a Work as this, and then sell the Impression for 1600 *l.* (as appears by their own Oaths) which Impression alone over-payes them all the Moneys they are out of Purse: And had they not been stopt in their full Carrear at the Council-Board, or rather by *Injunction* in *Chancery* (which they Complain of as a hinderance to their Trade also by Oath) they had by this time altered the Ancient Law-Books, and cast them into a new Modell of their own Invention; that by degrees the state and truth of the good Old Lawes by which Men hold their Lives and Estates, should utterly be lost and forgotten, and new Laws fram'd to fit the Humours of a new Invented *Government;* which they little value, so they may have full rates for their Books, and their Goddess *Diana*[36] be safe.

I have gone thus far upon mine own Strength onely, without any publique or private assistance; and because I am not willing to endure the Field much longer of my self, I think it my Duty to state the Case truly as it is, and implore the Ayd of such Neighbours (who cannot probably prevent the burning down of their own Houses, when mine is first set on fire): Common Experience tells us, a just Cause signifies little of it self, if it be not backt with Diligence and Friends; *Bonum apparens et bonum verum et absolute,*[37] are so like, (though of a Contrary Nature) that the Credit of the best Testimony gives either precedency; 'tis therefore not only hard, but impossible for one Man to Contend with a Thousand, and not be Conquered. *Hercules* was most Strong and Valiant, and yet, *nec Hercules contra duos.*[38] I have no proper Refuge but to his Majestie in this Case, which I do chiefly Espouse for his sake, who like King *David* is worth Ten Thousand of us. I have not the Power to *Impose,* but rather to *Propose* his Majesties timely Assistance: Onely this I hope I may say without offence, That if the King suppose it not for his Interest, I shall more willingly lay down the Cudgells, than I took them up at first: But if otherwise, I am as Careless of their Malice, as *Gallio,*[39] it being not the first time I have past upon the Forlorne Hope.[40]

'Tis against the Stationers Interest to redress the Evils of the Press.

By this time the Impartiall Reader may [be] inclin'd to believe, the Company of *Stationers* are not the fittest Persons to redress the Evills of the Press; because 'tis most certain, that none but themselves can offend: And 'twould be greater Self-denyall in them than can be expected, to punish themselves contrary to their Interests. There are at least 600 Booksellers that keep Shops in and about *London,* and Two or three Thousand free of the Company of *Stationers;* the Licensed Books of the Kingdome cannot imploy one third part of them: What shall the rest do? I have heard some of them openly at the *Committee* of the *House of Commons* say, They will rather hang than starve; and that a man is not hang'd for stealing but being taken; *necessitas cogit ad turpia.*[41] But this is not all, 'tis not onely for their Interest not to Suppress them, but to Maintain them: An unlicensed Book bears Treble the price of another; and generally the

THE ORIGINAL AND GROWTH OF PRINTING

more Scandalous a Book is, by so much the more dear: This hath inricht the Wealthiest of them; and when they fear losing their ill-got Goods, they put their Journey-men of the poorer sort, or their Apprentices, upon the Work, going shares with them, and taking their Oaths or other Security to be true to them, though false to all the World besides. Thus do they breed up their Youth like the *Lacedemonians,* who allow'd their Children little or nothing but what they could get by filching and stealing:[42] That the *Printers* are Poor and Numerous, can no body deny, for it hath lately been the great Work of this *Parliament* to lessen their Number, and to provide for their Poor. But because Extream Prices may be doubted by those that do not usually buy, I will give you one Instance for all; I was lately in a Book-seller's Shop, where I saw a Book in *Quarto,* entituled, *Killing no Murder,*[43] it had but eight Leaves in all, stitcht up without binding, he demanded 5 *s.* for it, and would not take less: A Book of the same bigness Licensed, would have cost but 4 *d.* or 6 *d.* at the most. 'Tis not then the Interest of the Company of *Stationers* to suppress unlicensed Books: Whose is it then? I Confidently Affirm, it is the Patentees, who derive from the King. I shall give you one Similitude of the like Nature; The King, as belonging to the Honor of *Windsor,* hath a great Quantity of Ground of which he makes little or no benefit, because it lyes in Common; And the Neighbours thereabouts, do not onely eat the Herbage, but steal the Kings Deer, and destroy his Woods, without giving any Accompt or Satisfaction whatsoever: To prevent which Mischief, the King Incloseth several Parks, and gives the keeping of them to Several Persons by *Patent,* reserving what he pleaseth out of them, the rest he gives the *Patentees:* these are still the Kings Parks, though kept by the *Patentees;* for the King kills what Deer he pleaseth, disposes of the Venison, and fells the Timber for Repair of his Houses, Shipping, *&c.* What wrong doth He to His Neighbours, by Inclosing His own Lands, which He denyes to none of His Subjects? Yet His Neighbours are troubled, because they cannot wrong Him as before; and upon every Distemper of

The unconscionable dealing of Book-Sellers.

RICHARD ATKYNS ESQUIRE

the Commonwealth, destroy the Fences, and make it Common again. Is it not (think You) the Interest of the *Pattentees,* to defend the Kings Right, and their own Profit under Him, and to prosecute the Law against such Offenders? Is not the Interest of the King and His *Patentees* so involved, that they cannot be divided? Just so is it by Inclosing 𝕻𝖗𝖎𝖓𝖙𝖎𝖓𝖌; the King (having the Right thereof, as much as of any Crown-Lands) Grants all sorts of Books, necessary for the Education of Youth, or the Improvements of Age, or whatsoever else is usefull for Soul, Body, or Estate, to several Persons by *Patent;* it will much concern these *Patentees,* in Honour and profit both, to see their several Grants be not Trespassed upon, nor Corrupted by others: And so they have ever kept their Copies intire, till the latter end of the late King *Charles* His Reign; At which time, the Company of *Stationers,* by fair Pretences, (as they did before to Queen *Mary,* to get their *Charter*) obtain'd a Decree of *Star-Chamber,* to Invest the Executive Power of 𝕻𝖗𝖎𝖓𝖙𝖎𝖓𝖌 in them, against the *Patentees;* and then Libellous and scandalous Books and Pamphlets began to fly about like Lightning: And when this was strengthned with an Act of 17 *Car.* which took away the Jurisdiction of the *Councel-Board* and the *Star-Chamber* (at least between Party and Party), their Mischief was compleated, and they impowred to vent the Passions of all Discontented Minds in Print, against 𝕸𝖔𝖓𝖆𝖗𝖈𝖍𝖞 and 𝕰𝖕𝖎𝖘𝖈𝖔𝖕𝖆𝖈𝖞; which they persu'd with such Diligence and Success, that they did eradicate both Root and Branch: Whereas the *Council-Board,* and the *Star-Chamber,* did usually Fine and Imprison such Transgressors; Of which, I can shew several Presidents, from the time of Queen *Elizabeth,* to this King's Reign.

Crown-Lands and Printing, equally the Kings Right.

The sad Effects of the Executive Power of Printing in the Company of Stationers.

Object. 4. But it may be said, *Scandalous, Libellous, and Heretical Books may be Printed of themselves, without any Relation to the several kinds of Books granted by Patent.*

Answ. I Answer, 'Tis very true; but as long as the Gospel, the Lawes, and all other Books for the Advancement of Learning, good Manners, and Education of Youth, that are in Grant, be kept intire, without any mixture of Heresie, Scandall or Schisme, 'twill go a great way in preventing Libells and Scandalls; and the **Parliament** may do the rest with greater Ease, by reducing the Number of **Presses**, and inflicting great Penalties upon such as shall **Print** and **Publish** unlicensed Books and Pamphlets: Which **Penalties** cannot be too big, because it lyes in their own power whether they will offend or not.

Too great Penalties cannot be inflicted for Offences in Printing.

If the Power this **Parliament** hath given the Company of **Stationers**, had taken any good Effect, they might have possibly continued it: But as a Worthy Gentleman Notes, amongst other most true, and Ingenious **Observations**, That not one Person hath been Fin'd, and but one prosecuted, by the Company of **Stationers**, since the late A C T, notwithstanding so much **Treason** and **Sedition** Printed and disperst since that time; but he himself (being an Active Gentleman and Loyall Subject) hath Discovered more within this two years, than the Company of **Stationers** have done ever since they had a **Charter**. To this give me leave to adde, that 'tis not likely (setting profit aside) even in order to Kingly Government, they should suppress these Books; for a **Corporation** being in it self a Petit-**State**, is inconsistent with **Monarchy**. Wofull Experience tells us, That very few, if any, went further with the K I N G than their own Interests led them; Which puts me in mind of a Story of Queen E L I Z A B E T H, being at *Quinborough*[44] upon an Occasion, the Maior of the town brought her onwards of her Way so far, till the Queen desir'd him to return back again, saying, he had brought her far enough: To which he replyed, **Madam, I'le bring your Majestie as far as my Way lyes**. For his Worship, being a Landed Man, had a small Tenement about a mile further.

More Treason and Sedition discovered by a Gentleman in two years, than hath ever been by the Stationers.

I wish **Corporations** would do but as much as that, and not go out of their Way to destroy **Monarchy**; for I must needs confess, that shining Shooes and set Ruffs were very forward to sit in Judg-

[29]

RICHARD ATKYNS ESQUIRE

ment upon the late King's Party, for doing but their Duty to their PRINCE, which they themselves ought to have done.

Humane Laws subject not only to Imperfection but Death it self.

And here I might take occasion to say, That though the **Lawes** of GOD be infinite and everlasting, and fitted for all Times and Persons, yet the Lawes of Men are like themselves, finite, imperfect, and subject to Infirmity, and Death it self, as the makers are: Hence it comes, that so many Laws are repeal'd, and others made in their rooms; and hence it may come, that all lesser Governments under a **Monarchy** may by *misuse* be wholly taken away, or else abated; as was heretofore the **Barons** Power by their so often taking up against the KING; and the two Hundreds of *Dudson* and *Kings Barton* who were by this very **Parliament** taken out of the Jurisdiction of **Gloucester**,[45] though enjoy'd by them ever since the Times of *Richard* the Third, the Grant being judg'd unreasonable; our best **Lawes** and **Constitutions** by Age losing strength and vigour, as our Bodies do, either by the Crafty Evasions of the Offendors, or by the alteration of their Vices, or by the male-administration of Justice upon the Malefactors; for the just performance of which last, none have greater Encouragement than the reward of a good Conscience to fortifie them against the malice of those they punish; who though they Act according to their best Judgment, yet by reason of some doubtful penning of a Law, Offenders are also many times taken out of the hand of Justice, to the very great encouragement of the Delinquents, and discouragement of the Punishers; insomuch, as when *Twyn*[46] was lately arraigned for **Printing Treason**, he was condemned by the old Law, though there be a new one now Extant for that very purpose. And so I return to the **Stationers** again; where I find them very Sollicitous with the **Parliament** to Enlarge, or at least Confirm, this Power they have already, (resolving to have it by *Hook* or by *Crook*) and Promising all care and diligence for the future, if they shall be, once more trusted; saying, with **Absolom**, *O that I were made Judge in the Land,* &c. *That every Man might have Justice:* and what followes, but presently raising an Army against King *David*, though his own Father.

[30]

THE ORIGINAL AND GROWTH OF PRINTING

Let not the *Reader* conjecture I lay an Imputation upon every particular Member of the Company, (for there are too many of them that groan under the like Burthen, as I my self do,) but against the whole Corporation, as a Body Politique (especially as 'tis now governed.)

I have now shewed you the Practices of the Company of *Stationers,* and some particular Members thereof, against the *King,* and his *Patentees;* who, like Painted Sepulchres, appear Glorious without, but within are full of Rottenness and Corruption: I have also opened, tented, and sering'd the Sores, of their Body Politique, and tryed them to the quick (which I hope will not prove like the Touch of a Gall'd Horse-Back, to receive a Kick for my Labour and good will). But because I am not so good a Chirurgion, as to close and cure them again my self I shall implore, the help of *Parliament,* and shall most humbly Recommend them to their Cure together with these ensuing 𝔒𝔟𝔰𝔢𝔯𝔳𝔞𝔱𝔦𝔬𝔫𝔰 and 𝔓𝔯𝔬𝔭𝔬𝔰𝔞𝔩𝔰.

The Reasons inducing Queen *Mary* to Incorporate the *Stationers,* are expressed in her 𝔈𝔥𝔞𝔯𝔱𝔢𝔯 in these words, *Nos considerantes quod seditiosi et heretici Libri, Rithmi, &c. indies sunt editi, excuss. et impressi per diversas scandalosas, malitiosas, Schismatic. et heretic. personas non solum movend. Subditos et Ligeos nostros ad seditiones et inobedientias, contra nos, Coronam et dignitatem nostras, verum etiam ad maximas et detestabiles hereses contra fidem, &c. Et remedium congruum in hac parte providere Volentes; de gratia nostra speciali &c.*⁴⁷ The Queen Erects the Corporation with Powers and Trusts,

1. To make Lawes *pro securo regimine* of the Members of the Company.
2. To search for and seize Books Printed contrary to Law.

Observ. 1. The Erecting this 𝔈𝔬𝔯𝔭𝔬𝔯𝔞𝔱𝔦𝔬𝔫 hath not proved Remediall against the Mischiefs recited in the 𝔈𝔥𝔞𝔯𝔱𝔢𝔯; But she Queen was wholly deceived in the Design aymed at in passing the 𝔈𝔥𝔞𝔯𝔱𝔢𝔯.

Observations and Proposals recommended to the Parliament.

RICHARD ATKYNS ESQUIRE

2. The Intrusting the 𝕾𝖙𝖆𝖙𝖎𝖔𝖓𝖊𝖗𝖘 with the Powers aforesaid, hath not only not remedied, but hath encouraged, encreased, and secured the Printing [of] 𝕾𝖊𝖉𝖎𝖙𝖎𝖔𝖓 and 𝕿𝖗𝖊𝖆𝖘𝖔𝖓. For,

3. The Persons who are Intrusted with the Search and Discovery of the Offences to be remedied, are themselves the Common Offenders therein.

Principall and Particular Members of the Company high Delinquents.

The Company, in their Politique Capacity, cannot Exec[u]te the 𝕿𝖗𝖚𝖘𝖙𝖘, or merit or offend, but by their particular Members; divers principal Members of the Company have been actually Convicted, some as privy and accessory, other as Principals, in Printing and Publishing Illegal Books; and many Treasonable Books have been printed; during the late Troubles, for several principal Members of the Company.

4. The Company of *Stationers* have in other things exceeded the Authorities granted by their 𝕮𝖍𝖆𝖗𝖙𝖊𝖗, as by Imposing and Administring of Oaths, *&c.* and by Entring other Mens 𝕮𝖔𝖕𝖎𝖊𝖘 in their Hall-Book as their own, and then Printing and Selling them, in Opposition to the King's Grant; and this by vertue of a Law in the late Evil Times: and have also assumed to themselves (by Colour of the said 𝕮𝖍𝖆𝖗𝖙𝖊𝖗) the whole Right of Priviledging and Exercising Printing, and have Combined to oppose and overthrow the King's Just Power and Prerogative herein, and Interest of His *Patentees*.

1. From all which it appears, the Crown hath been deceived in the End and Design of Erecting the said 𝕮𝖔𝖗𝖕𝖔𝖗𝖆𝖙𝖎𝖔𝖓.

2. That they have not proved Remediall but Instrumental, to the Increase of the Mischiefs they should redress.

3. That they have broke and acted contrary to the 𝕿𝖗𝖚𝖘𝖙𝖘 imposed in them by their 𝕮𝖍𝖆𝖗𝖙𝖊𝖗.

4. They have by Colour of their 𝕮𝖍𝖆𝖗𝖙𝖊𝖗, abused the Favour of the Crown, in exceeding the Authorities granted them, and assuming to themselves the whole Power of the Crown, concerning the Matter of 𝕻𝖗𝖎𝖓𝖙𝖎𝖓𝖌.

I therefore take the boldness, most humbly to propose to your Honours;

I. That the King's Just Power and Prerogative, in the impowring and restraining **Printing**, and in the Hearing, Regulating, and Determining all Differences touching the same, as a Matter of State, be Declared and Confirmed, as an Antient and Hereditary Right of the C R O W N; And that all Laws contrary thereunto be Repealed.

The Proposals.

II. That an A C T for Regulating **Printing** may Establish Propriety therein according to the Kings Grants thereof, and may direct Rules for the Licensing and Management of **Printing**, and inflict Penalties for Abuses therein, with Legall Means for the Executing such **Penalties**, and for settling and securing every Man's **Propriety**, (saving the Right of the **Crown**) to regulate or restrain any Interest or Right in **Printing**, or other matter concerning the same, which by the **King** and **Councill** shall be conceiv'd a Nusance of State.

III. That the **Charter** of the Company of *Stationers,* who Claym thereby an unlimited Power in **Printing**, be examined, together with the **Unreasonablenesse** therof, and the **Abuses** committed thereby by Testimony of **Witnesses** to be Summoned to that purpose; And that the said *Charter* and the *Powers* thereby granted, be limited according to **Reason** and the true Intent of the **Grant**.

IV. That the **Penalty** for **Printing** without Licence, be forfeiture of the Book or thing so **Printed**, and treble the value thereof, one Moyety thereof to the **Patentee** or Party interessed in the Right of Printing such unlicensed Book (if any person be therein interessed), or otherwise to the K I N G, and the other Moyety to the **Informer**: But that Books once Licensed, may be **reprinted** without Licence, or so much of them as shall be without addition or alteration.

V. That the K I N G's **Patentee** for **Printing Law-Books**, be Priviledg'd with a like Priviledge, as the **Patentees** for the **Bible** are, or shall be Priviledged; and with Power to search with an Officer for **unlicensed Law-Books**, and to Seize and carry away the same to some publique place.

VI. That no Disloyall or Notorious Criminal Person for Printing **Treasonable** or **Seditious Books** in the late Times of Trouble, be admitted to keep a **Printing-Presse**; And that such as be Intrusted with a **Printing-Presse**, be Sworn not to offend the A C T of **Parliament**, &c. and give Security for the same.

VII. That the Entry of other Mens **Copies** in *Stationers*-Hall, be declared to be of no Validity, especially as to give them any Title to such Books as are Granted by **Patent** to others.

Object. 5.   *And now it may be most truly said, That the Author is very tedious, and yet hath made few or no Propositions but such as concern the King and his Patentees.*

Answer.   To which I Answer, That all other Interests have not been wanting to make the best of their Case, and their Desires to be fully understood; And as for the Company of *Stationers,* they were by this late A C T so amply provided for, as that at the **Committee** of the **House of Commons** they had nothing more of Substance to desire. The **Printers** have also published a late **Book**, wherein they desire to be Incorporated and made a Company of themselves, apart from the Company of **Stationers**, of which they now are; and therein also have stated the best of their Case, Mr. *L'Estrange*[48] hath also published a **Book**, wherein he Treateth the whole matter in generall, shews the severall **Abuses** of **Printing** and **Printers**, but hath not applyed himself to any particular Interest: And therefore I have taken the Boldness to say somewhat, though weakly, for the K I N G and his **Patentees**; hoping an ill **Pen** shall not destroy a good **Cause**; But that the **Wisedome** and **Loyalty** of this **Parliament**, which is Exemplary for both, will Supply all Defects, and take the Will for the Deed; The rather, because Extream Necessity enforceth me to say somewhat now before the A C T be past; Which makes me rather adventure to be ridiculous, than wanting to my Duty. I shall add onely one word more, That in a Business of so great **Intricacy** and **Concernment** as this of **Printing**, your Honours would not without very great Consideration, make an A C T for Perpetuity, In which all **Interests** may be equally Considered; the rather, because the late A C T now in being, which was past in hast, is judg'd (even by Your Selves) to have many **Imperfections** in it.

And if the Brewers, who at most can but steal away a Flegmatick part of the King's Revenue,[49] deserve the serious Consideration of the Supreme Council of *England*, how much more these, that do not onely bereave the King of his Good-Name, but of the very Hearts of His People; between whom there is as much oddes, as between a Pyrate that robs a Ship or two, and *Alexander* that robs the whole World.

FINIS.

THE
KINGS
Grant of Privilege
FOR
SOLE PRINTING
Common-Lavv-Books,
DEFENDED;
AND THE
Legality thereof
ASSERTED.

RICHMOND:
Printed for Tiger of the Stripe, 2013

THE
KINGS Grant of PRIVILEDGE
FOR
Sole Printing *Common-Law-Books*, defended, &c.

The principal Exception against the Grant of Priviledge before mentioned is the slander of a Monopoly; And the principall Foundation on which it stands supported and justified is the Kings Prerogative; and therefore those two matters, A Monopoly and the nature thereof; and the Kings Prerogative and the Extent thereof are necessarily to be first considered.

I consider a Monopoly as it is (or was) at the Common Law before the Stat. 21 Jac. the matter in question being of a Grant made before that Statute; and the thing granted Excepted out of that Statute.

'Tis true, A Monopoly is (as many other ungrateful terms are) taken primarily and generally in the worst sense, to signifie something unlawful, against the Freedome of Trade, the Liberty of the Subject, &c. And the word is thereupon also forced and extended (beyond its literal signification) to comprehend every sole dealing or exercising of that which others are restrained to use. And to be termed a Monopoly, is at this day an imputation, as if the unlawfulness thereof were necessarily to be presumed and implyed. Whereas it is most clear, That All Monopolies are not against Law, some being reasonable, useful and beneficial to the Publick, and some necessary; and this Necessity and Benefit to the People recompenceth the restraint of their Liberty, and taketh away the unlawfulness thereof. All Patents of Priviledge for the sole usage of new inventions are Monopolies undeniably, yet is it necessary they should be granted for the encouragement of Industry and Invention; The communicating whereof to public use, is a public and general benefit, though the making the Invention should be perpetually appropriated to the Inventor. All or most of the Ancient Offices were at the first (and are agreed to be) direct and plain Monopolies in their natures, and are now found so to be in their mischievous consequence to the generality of the people, whose charges do but enrich a single person sometimes for performing that which is needless; or if needful, might be performed by the persons concerned themselves without charge. Nevertheless, such Offices having been created originally upon reason of Use and Benefit to the Publick, for encouragement of Learning, Diligence or Fidelity, or such like Motives to His Majesties Royal Predecessors by whom they were created, have from Age to Age been approved and continued as they now are, (and are excepted by the name with this Priviledge in question out of the aforesaid Stat. 21 Jac.) as unquestionably lawful.

It hath been said, That Ancient Offices are established and made Lawfull by Time and Custome, which is part of the L A W. To which I Answer, That a Monopoly is an Evill of that nature as could not be justified by Custome, or by length of Time, if it were not ex rationabile causa usitata,[50] a benefit to the People in recompence of the restraint of their Freedoms: For 'tis the reasonableness of Benefit that justifies the thing, and not Time or Custome; In Consuetudinibus consideranda est soliditas rationis, non diuturnitas temporis.[51] In like manner a man may by Usage or Reservation claim the Sole Priviledg of keeping a common Mill, a common Bake-house or Brew-house within a certain Precinct; For this may have commenced originally ex rationabili causa, by bargain or agreement to be made at the Owners charge, and for the Inhabitants ease and benefit, which is a recompence for the restraint of their liberty of using the like.

2 E.3.7. The Case there is, That the King had granted a Charter of Priviledge to the Lord or Owner of a Haven, that such Ships as Anchored or entred there for Harbour, should unlade there only. This and the precedent Cases are plain Monopolies; yet because they stand upon equivalent benefit, and the Ships had harbour

THE KING'S GRANT OF PRIVILEGE FOR SOLE PRINTING

and safety from the Lord of the Haven, 'twas therefore allowed a lawful Charter.

From which Cases, I infer, That before the Statute a Monopoly might be lawfully erected, because it might be beneficial to the Publick, and was permitted in special Cases. And with this agreeth the Learned Grotius in his Book de Jure Belli & Pacis.[52] Monopolia non omnia cum jure naturæ pugnant, nam possunt interdum à summa potestate permitti justa de causa, &c.[53] And he instances the practice and permission thereof in the Roman State, (the pattern of Governments) and the Holy Story of Joseph touching this matter. A Monopoly is then unlawful, when thereby the People are restrained in their Lawfull Trades, or in the exercise of what they have right to use, without general benefit or recompence for the same: but the priviledging particular persons to exercise a particular Imployment which others never did use, nor have the right to use, and the generall use whereof would be dangerous, and the Restraint of the use safe for the Publick, cannot be unlawful; for the reason of that unlawfulness fails, cessante ratione Legis cessat Lex.[54]

Lib. 2. cap. 12. Parag. 16

Now for the Prerogative (which is a copious subject) I shall only mention so much touching the same, as I conceive most proper to the matter in hand. The King hath Prerogatives of severall natures, and grounded upon severall Reasons; Some in respect of his own Royal Estate and Person; Others in respect of His Office and Magistracy, for the better Government of his People: For the King as Supream Magistrate hath a general Trust and Care of the Peoples safety, to prevent, as well as to deliver from public Evils. Rex &c. ratione dignitatis Regiæ ad providendum salvationi Regni circumquaq; est astrictus.[55] Now Providentia is (Properly) futuorum; whereby the King is to use all means to foresee and prevent mischiefs within his Kingdom. For this purpose, and for the enabling him to perform this Office and Trust, he is by Law endowed with several transcendent Prerogatives, some known, and some unlimited and unknown which jura sum-

Reg. 127.

[41]

RICHARD ATKYNS ESQUIRE

mi Magistratus, **as great for weight, and as infinite for number as the contingencies may be wherewith the Peoples safety may be affected.**

The Extent of the Kings Prerogatives, **such as concern their own personal rights, or the rights of their Estate, are sometimes disputed, and have sometimes been limited and restrained by their own consents in Parliament. But those touching the preservation of the publick have never been limited, nor ought to be disputed or lesned; and if so, the intended limitations and restraints thereof have been adjudged void because these** Prerogatives **are inseparable from the Crown.**

Hence it is, The King can dispence with Laws, can pardon offences, can licence matters prohibited, can prohibit matters tolerated, and can priviledge, restrain, or qualifie new accidents, as he in Wisdome and Deliberation shall judge expedient and best for the Publick good. Which Judgment and Deliberation is peculiar and proper to the K I N G, who alone comprehendeth the Estate of publick things, **and it is a duty and a consequence of his Supream Magistracy. Now** Printing **in every mans reason and observation is, and in the Act for Regulating** Printing **is prefaced to be matter of** Publick Care **and** great Concernment.

F. Paul Servita in his History of the Inqu. *pag.* 104.[56]

These things being premised, I shall only state the Case truly as it is understood touching the Priviledge **in question, and then the Application will be obvious.**

Anno 1466.

In the Reign of King Hen. 6. **the** Art of Printing **was first invented. And as some** Manuscripts **relate, the same King** Hen. 6. **purchased the first discovery of the Act, and thereby became Proprietor thereof at his own charge; Whereby the same came to be taught and used in** England, **but for the Printing of such matters onely as the King** Licensed **and** Privileged, **and by the sworn Servants of the King onely, and in places appointed by the King, and not elsewhere.**

This appears by a Manuscript thereof in the Library at *Lambeth.*

[42]

THE KING'S GRANT OF PRIVILEGE FOR SOLE PRINTING

By the later end of the Reign of H. 8. the Invention was come to some perfection, and several Books were then printed here cum Privilegio, and Others brought over printed from beyond the Seas; but being few in number, and the prices thereof excessive, the same was remedied by the Stat. 25. H. 8. cap. 15.[57]

The State at that time taking consideration of the growth of Printing, and the danger and consequences that might ensue to the King and People by Printing the Lawes of the Land, that thereby Errors and Seditions might be divulged and insinuated, and other mischiefs happen to affect the people, thought fit thereupon to commit the Printing of the L A W E S to the Care and Trust of some particular persons whom the K I N G by Patent priviledged to print the same, with a Clause of Restraint to all others from presuming to meddle therein.   *7. E. 6.*

All succeeding Kings and Queens of this Realm have upon like Considerations in their Grants, and other Considerations of State, in Wisdome thought fit to continue the said Priviledg in the hands of some persons in whom they confided, with like Clauses of restraint as before. The dates and successions of which Grants are as followeth.

The King granted a Patent of Sole Priviledge to Print Law-Books, to Tottell for 7. years, with restraint to others from presuming to print his Lawes.   *7. E. 6.*

The Queen renewed Tottells Grant for life.   *1. Eliz.*

The Queen granted like Patent of priviledg to Yestweirt Clerk of her Signet, for 30 years.   *20. Eliz.*

The Queen granted a New Patent of like Priviledg to Weight and Norton for 30 years.   *41. Eliz.*

The King granted a New Patent of like Priviledg to John Moore Clerk of his Signet for 40 years; Which Patent is still continuing.   *15. Jac.*

These Priviledges ad Imprimendum solum, have continually been enjoyed according the the purport of the said Grants; saving the interruption forced upon the Presse after 1642. in the times of the late Troubles, whereby Sedition and Treason came to be printed openly, and continued so to be till his Majesties Restauration.

This is the first peaceable Age wherein the Kings Prerogative, in this matter of Printing the L A W E S , was ever questioned, or the aforesaid Priviledges charged with the imputation of Monopolies. And whether they be Such Monopolies as are against Law, is the present question.

For the justifying the lawfulness of this Priviledge, I offer the Reasons ensuing.

1. That the King hath as absolute Power to prevent Evils foreseen, as he hath to reform them which happen unforeseen. And I conceive it clear, as He may forbid the exercise of any Invention, which upon the Permission thereof shall prove or become a Nusance, or common mischief, So He may qualifie, or wholly prohibit the first use of it, out of a prospect of the mischief. Watchfulness and Carefulness are the Duties required of a good Prince; to watch, is, that He may prevent and obviate Dangers. Now experience hath discovered to us the Dangers and Mischiefs of the Liberty of Printing; And, though the excellency of the Invention cannot be denyed, yet, whoever will consider it, shall find, that Factions and Errors in matters of Religion, and principles of Treason and Rebellions in matters of State have been more insinuated and fomented by the Liberty of the Presse, then by any other single means. So it may seem a question (impatiently considered) whether the Use of Printing recompenceth the mischief by the liberty and abuse thereof. Therefore the [a]Father observeth excellently well, The matter of Books seemeth to be a thing of small moment, because it treats of words; but through these words, come Opinions into the World, which cause Partialities, Seditions, and Wars: They are words, it is true, but such as in conseqence, draw after them Hosts of Armed Men.

Now certainly, had the King at the first discovery of the Invention of Printing, foreseen the Use thereof a likely means of

a F. Paul
Servita ubi supra, Pag. 106, 107.

[44]

Disturbance to the peace of the Church or State (as the liberty and abuse thereof hath proved to both), it had been in the Kings power, for the Peace and Safety of both, to have prohibited the Use of Printing wholly.[58]

2. As upon the Reason aforesaid, the King might at the first have refused to have received the Use of Printing at all into His Dominions, so much more reasonably might He restrain the general Liberty and Use thereof, not to extend to matters of State or Law, these being peculiarly within His Concerns, and of more apparent danger to the peace of the State. Some States have not suffered their Laws at all to be published or known. There might be policy in this, though it seems unjust; yet on the contrary, for a general Liberty to publish is neither honourable nor safe. The mean betwixt these extreams hath been practised by the Kings of this Realm, not to Restrain the printing of the Laws wholly (as They might have done) nor yet to give a General Liberty to every man for the doing thereof (which might prove unsafe) but to Priviledge Select Persons only to do the same, who might be answerable for Misdemeanors and Defects therein.

3. Though the Art of Printing was discovered sometime before the Reign of E.6. from whom the first Patent of this Priviledge appears Granted, yet were the Presses all then Licenced by the King; And none, or no considerable Book of the Law was printed before that time, the Act not having come to perfection: So that the first Patent of this priviledge could not be pretended a Monopoly, or illegal, none then having the Trade, or Right of Printing the Laws to be detrimented thereby.

4. The King having at the first beginning of Printing by his lawful Prerogative, and upon just Reason placed this Priviledge of printing the Laws solely in the hands of particular persons, to prevent mischiefs which might ensue upon a general liberty given to print the Laws; and the said Priviledge being then not unlawful, because no restraint of any thing then practised or exercised, or which any one had right to exercise: and having ever

since so continued, and the people generally neither intitled to the Right or Usage of printing the Laws, remains grantable as at the first by Virtue of the same Prerogative, and for the same ends, and with the same innocence from injuring any one.

5. Besides the Reasons before mentioned, (of security to the Kingdome, against Innovations, or false construction of the Laws, either by the designs of Authors, or mistake of Printers, which is worthy the Princes Care, and those He entrusts with the printing of the Laws to prevent) the King hath (as I conceive) a peculiar Right and Property **(not only in the Act and Invention of Printing by purchase (as before hath been said) for in that I lay no great weight but** in the Laws **themselves, and in the** publishing **thereof, which cannot be taken from Him, or assumed by any Subject without His leave.** 'Tis true the people have also a right in the Laws (as they had before Printing was known) not to print them, but to receive the fruit of them from the Kings Hand. But the King is the Repository and the Proprietor, and is entrusted with the Promulgation and Execution of the Laws.

There is Lex scripta & non scripta. **The Written Laws are Records,** &c. **which are** Recorda & Brevia Domini Regis, **and are reckoned** inter Thesauraria Regis,[59] **as the chief and principal things wherein he hath property. But the unwritten Laws, which are grounded upon Custom and Reason,** &c. **are yet more properly the Kings then the other, for these are in his Brest. The Written Law is reposited** in Arca **or** Thesauro Regis, **but the Laws unwritten are** in pectore Regis. In scrinio pectoris,[60] **saith** Fortescue, **From whence I infer, That these Laws and Records which are so peculiarly the Kings in property and dispensation, ought not to be published, or numbred, or interpreted but by Authority from Him; and the printing thereof is of the Kings free pleasure, and not the peoples Right, and consequently the priviledging some to print the Law is the Kings Grace, and the restraining others from that Liberty not any wrong.**

Cap. 45.

6. If no material Reason could be offered in this Case, to assert the Kings Right in Granting this Priviledge, yet there want not Authorities to justify the same.

1. The constant usage and practise, without exception from the first settlement of Printing, as appears by the succession of Patents before mentioned.

In the Argument of Darcy and Allens Case,[61] One great Reason against the Patent, was, That the like had never been granted before. But here the like hath ever been granted, ever since the printing of the Laws, and the like (or any) Exception thereunto never heard of before.

44. Eliz.

2. The general Allowance of the Judges in the Argument of Darcy and Allens Case, where this Patent was cited as a president, and holden lawful, & necessary pur le peace & safety del Realm, *Nemine contradicente.*

Moores Rep. 673.[62]

3. The Stat. 21 Jac. cap. 3. was passed purposely to suppresse the then present, and to prevent the future granting of Monopolies, and yet expressly excepts Patents of priviledge for sole printing of Books with several branches of the Militia and Offices, and other like things of the highest concern to the Crown. And I cannot omit to observe, That this priviledge of printing is the first thing named in the Exception, as if the Parliament then had it first and principally in their Care; And that this Patent now in question was the same Patent then in force.

4. The Stat. 21 Jac. before mentioned, and also the Stat. 14 Car. 2 touching the Regulation of printing, provide for Patents of Priviledge for printing, Granted, or to be granted; which they would not have done, had they not approved and intended to encourage like Grants to be made. And also the last mentioned Stat. fol. 433. expressly prohibits under penalties printing or com-printing of such Books, the which any other hath sole priviledge to print by Letters Patents; which implies, the Parliament intended to support and establish such as lawfull; and it cannot be reasonably thought several Parliaments should so expressly provide

for this priviledge of sole Printing, if they had not designed to secure it from the Censure of a Monopoly.

1. Object.

It hath been Objected, That this Patent hath the mischiefs of a Monopoly, for thereby the Patentee may enhaunce the prices of Law-Books; May print the Law-Books as defectively as he pleases, and may prise mens Labours at his own Rates, &c.

Sol.

1. The Prices of Books may (if occasion shall be) be regulated by the Chancellor, &c. per Stat. 25 H. 8. Cap. 15.

2. Defective printing, or other abuses in or about the printing of the Laws, is a breach of the Trust, and punishable in the Patentee, and a cause of forfeiture of his Patent, as mis-execution, or defective execution is a cause of forfeiture of an Office.

3. If these Objections were true, and could receive no Answer, the Mischiefs pretended are not comparable to the benefits received, or the security which redounds to the Publick, by restraining the general Liberty of printing the Laws.

2. Obj.

The words of the Patent are said to be too large and unreasonable, to priviledge all Books concerning the Common Laws. For herein all manner of Books whatever are included, forasmuch as every Book more or less compriseth something of the Common-Law.

Sol.

This is an unreasonable construction of the words; for Books treating principally of another Subject, which in the proof, or proceeding thereunto, only mention some Maxims or principles of the Law, can only be said to contain in them some Chapter or Page (but cannot be termed Books) concerning the Law, the Law neither being their subject or design. Denominatio sumitur à Principali.[63]

7. If this Patent touching the sole printing of the Laws should in this Age have the sentence of a Monopoly against Law, in consequence other Patents or priviledge of like nature for sole printing of Books (that is to say) the Patents to the Kings Printers for printing Proclamations. The Patents for printing the Bible, Testament, Common-Prayer, &c. The Patents of both the

[48]

THE KING'S GRANT OF PRIVILEGE FOR SOLE PRINTING

Universities of this Kingdom in reference to Printing. **The several Patents to the** Company **of** Stationers **for the sole printing of the** Primer, Psalters, Singing Psalms, School-Books; **and that of** Almanacs, **the words of which are,** All and all manner of Almanacks, *In terminis* **such as be the words of the Grant in question, and are all Priviledges of the like nature and authority (but of lesse reason and use) must have the same fate to be overthrown therewith.**

8. **The usage of other Neighbour Kingdomes and States, may in this matter enforce the reasonableness of the like Usage here. In** France, Germany, Holland, &c. **sole priviledges of this Nature are usually granted, and solemnly observed. The Forms whereof are to be seen before several Books printed within those Kingdomes; to this effect,** (*viz.*) Sancta Cæsarea Majestas diplomate suo sanxit, ne quis præter A. B. C. D. intra sacri Imperii Romani Regnorumque, &c. Fines, **these and those Books** in toto vel in parte excudat, &c. sub confiscatione, &c.

In like manner, Ordinum *Hollandiæ, Westfrisiæque* singulari privilegio cautum est, ne quis præter A. B. & C. D. **(these and those Books)** Imprimat &c. sub confiscatione, &c.

The Form of the French Kings Priviledges, recite his Prerogative, That no Book can be printed within his Dominions, Sans avoir sur ceo nos Lettres à ce neccessairs, **That thereupon he does permit such persons to print such Books in such manner,** &c. Faysans d'offence à tous imprimeurs, &c. d'imprimer. &c: **any of the said Books besides the persons priviledged.**

Now forasmuch as the Kings of this Realm of England are not restrained herein (in case they might so be) by any Statute since the Invention of Printing, why should They be conceived to have less Right and power to Grant like priviledges touching printing, then is practised by their Neighbour Princes upon the same Reasons of Law and State, for their Subjects safety. It being almost impossible for a Prince to rule the spirits and wills of his Subjects (since Printing came in Use) without restraining the Presse, which so evidently influences to Evil or Good. I

only add, That after the Long Parliament had (Anno 1641.) opened a Liberty to the Presse for their first Service, to insinuate and propagate Principles of Rebellion, they immediately found it necessary again to Restrain the same (Anno 1643.) for their own Security.

The sum of all which is,

1. That some Monopolies may be necessary and useful to the Publick, and consequently lawful,

2. That the King hath Prerogative to priviledge such, and is Judge of the matter.

3. That the Priviledge in question is such, and hath been so adjudged by the Kings Predecessors ever since the Reign of Edw. 6.

4. That there hath been a continued succession of Patents of the same priviledge ever since the printing of the Laws.

5. That experience hath discovered the Mischief of Liberty in printing the Laws.

6. That the King upon fore-sight hereof (much more upon experience) might Restrain the printing of the Laws wholly.

7. That the King hath a Property in the Laws, and 'tis His Grace, and not the peoples Right, to have them printed.

8. That like priviledges for sole printing of Books, are practised and used to be granted by all the Neighbour Princes and States where printing is used.

9. That in Arguments of Law, this Priviledge hath been cited and allowed lawful.

10. That several Statutes have excepted and preserved it as lawful. From all which it is (with submission) concluded to be so.

FINIS.

THE VINDICATION OF Richard Atkyns
ESQUIRE

As also a Relation of several PASSAGES
in the *WESTERN-WAR*,
Wherein He was Concern'd

Together,
With certain SIGHS or EJACULATIONS
at the End of every *CHAPTER*.

Dedicated to His particular FRIENDS:
And intended to no other

Qui jacet in terris non habet unde cadat.

RICHMOND:
Printed for TIGER OF THE STRIPE,
MMXIII

The Preface.

*Do not find many People the better for* Vindications, *and as few as may be shall be the worse for this of mine, having confin'd myself, in the Application of it, only to Persons concern'd, that the trouble of it may not extend further than the Necessity: I call it a Necessity, for I have only this choice left me, either to suffer in my Name, Family, Credit, and in truth in my entire Character, and the whole Story of my Conversation, even among my most familiar Acquaintances and Allies, or else to redeem my self by this Defence; which I have also so digested, as to answer the several Parts of the* Calumny. *The* Scandal *it self is a very great Affliction, but the Hand from whom it comes, is yet a further aggravation of it. I had it in my thoughts to have acquitted my self in Writing, but finding the Task too much for a Pen, I was enforced to make use of the Press, keeping however to my first Intention, of placing the Copies only in those Hands where my Adversaries have employed their* Malice *to my* Reproach.

If perchance any of these Papers shall happen to be otherwise disposed of, let the Reader stop here, or blame himself rather than me, for a trouble I never meant him.

But unto such as they shall properly come, I do declare, I found it necessary to bring in the most considerable Passages of my whole Life, without which, my Vindication *alone would have seemed imperfect and abrupt, the Causes of it relating to more than Thirty Years since, and by which the Reader may judge, whether it be probable that such Premises should produce such Conclusions as are most untruly laid to my Charge.*

The Vindication of Richard Atkins Esq;

I Have heard, that when the Pope and Cardinals of Rome send their Emissaries into any of the Reform'd Churches, they injoyn them to Write their Lives; by which they may ground a Judgment upon the nature and disposition of the People to whom they are sent, as well as the dispatches of those whom they Send: And though I shall never encourage so much nonsense in the World, as for every man to write his Life; yet I shall never judge any man that doth it, because every Man is an Emissary from God; and if one Hair (which is an excrement) doth not fall without the providence of an Almighty Power, how much more providence may be expected upon the whole man, who is Gods Image? And if the well drawn Pictures of our Auncestors adorn our Houses, how much more should the Painting their own minds adorn our Lives and Conversations. If they have done well, to stir us up to outdo them; if ill, to give us cautions to avoid the evils of them. I cannot therefore condemn writing in any, because no man lives so unnecessarily but that God is glorified in his life; Yet I should not approve of it in my self, but that necessity compels me to countermine the deep laid Plots and Designs contrived against me, by my false, and active Enemies, who study to render me the worst of men: And if *a good name is better than pretious oyntment,* an ill name will stink to all posterity: That is it hath stirred me up to run this hazard of censure; with this also, that my Name and Family on both sides, might not suffer reproach through the consent of my silence.

Eccles. Cap. 17.

My birth was neither glorious nor contemptible, descending from Gentry on my Fathers side, and Nobility on my Mothers; my Father being Son and Heir of *Richard Atkyns* Esq;[64] of *Tuffleigh* in the County of *Gloucester,* Chief Justice of *West-Wales,* and of Queen

RICHARD ATKYNS ESQUIRE

Elizabeth's Councel, in the Councel of the *Marches* of *Wales;* and Brother to Sir *Edward Atkyns* of *Lincolnes-inn,* one of the Barons of the Exchequer. My Mother second Daughter of Sir *Edwin Sandys* Banneret of *Lattimers* in the County of *Bucks,*[65] by *Elizabeth* Lady *Sandys,* Daughter and Heir of *William* Lord *Sands* of the *Vine* in *Hantshire,* begotten of the Body of the Daughter and Heir of the Lord *Bray.*[66]

No sooner had I life, then I was like to lose my Lively-hood, but by Gods providence, and the assistance of Friends that Cloud was soon over. Whilst I was at Nurse my Arm was broke, and being ill set, I was conveyed to London (being above Fourscore Miles from the place) to have it set better; where playing amongst Horses in *Covent-Garden,* one struck the Skin off my Forehead with his hinder Feet, and did me no further mischief.

1st Sigh.	*— Lord, thou art he that took me out of my Mothers Womb! thou wast my hope when I hanged yet on my Mothers Breast: But Lord, what dangers and mischiefs is Man subject to! There is but one way to Live, and hundreds of ways to be depriv'd of Life.*
My usage at School.	After I was taken from the tuition of Women, and capable of going to School, I was plac'd under a Knights Chaplain (my Mothers Uncle) who Taught none but his own Sons, and my self; the Chaplain being Young and Amorous, made Love to one of the Chamber-maids, and Scor'd every frown she gave him on my Breech; his Patrons Sons were big enough to remedy the mischief, or to complain of it; so all the stripes happened upon me, that for the space of one whole Year, I was whipt most dayes twice a day, and the buds of the Rod so thick planted in my Skin, that had it been a Fruitful Soyl, it must needs have produce a whole Coppice of Birch; of which I durst not complain, for fear of being whipt for complaining also; But a discovery being made, and the truth of the matter examined, I was sent to the Minister of the Parish to School, who
Jealousy void of reason.	was Jealous of his Wife; and upon every dispute between them, would (not take the pains to whip me, but) take me by the Hair of

the Head and beat it against the wall, which caus'd me to be very dull; and giving me Lectures much more than I could digest, I was so heartless, that I benefited but little in two Years time. After this I was sent to a Free-school in *Gloucester*,⁶⁷ where I made Exercises best of any in the Seat; but repeated Grammar Rules and Lessons worst, by reason of the usage aforesaid; so as my Master suspected others to have made my Exercises for me, which took me off from making them so well as I could: But when I began to take some delight in Learning, I fell sick of the Small Pox, and was very near Death, and went to School no more.

— Lord, what variety of troubles are incident to the nature of Man! and if these be the portion of our innocent age, what reward must we expect for our riper and more sinful Age!   *2d. Sigh.*

I was afterwards sent to the University of *Oxon*⁶⁸ about fourteen years of Age, where I may too truely say I spent two years time; what by not being so well grounded as I ought to read a *Greek* or *Latin* Author with pleasure; and being placed under a Tutor, who (though he were a compleat Gentleman, and a Scholar, yet) by reason of his much absence, (being in a wooing condition) read very Seldom to me; so as I frequented Publick Exercises little, and studyed *Zegardine's* Philosophy⁶⁹ more than *Aristotle,* or *Ramus,* thereby imitating the *Sicilians* to erect a Temple to Riot; But I discovered, and resolv'd to amend this Error. When I was taken from thence,'twas debated among my Friends, whether to send me to the Inns of Court, or to Travel; and being the only Child of my Parents, the first was resolved upon: In prosecution of which, I was sent to *Lincolnes-Inn* (the place where the Family of my Fathers side had anciently been, and some of them then there) but upon some disgust in my entrance, I was again recalled from thence to the *Vine,*⁷⁰ where my Grandmother (the Lord *Sands* Daughter, and Heir) then livd: I had not been long there, but the Lord *Arundel* of *Wardour*⁷¹ came thither upon a Visit, and brought his only Son by the Second *Venter*⁷² with him, who was design'd for Travel; we be-   *Resolution for Travel.*

At the University.

RICHARD ATKYNS ESQUIRE

ing much of an age and humour, soon struck up an acquaintance and an agreement to go together; there was but one obstacle in the way, (*viz.*) that there was great danger of changing my Religion; his Tutor being a *Romish* Priest, and his Lordship having been a great Defender of the *Romish* Faith; not only by his Sword in taking the *Turks* Standard, for which he was made a *German* Count,[73] but by writing for the Church of *Rome*, against Bishop *Jewel*;[74] and as there was danger on the one hand as to matters of Faith, so there was safety of person and conveniency on the other: His Lordship giving us Letters of credence to all Christendome, being of greater renown abroad than in his own native Country; by which means, and the intimacy I had with my Cosin *Arundell*, I suppose I saw those things that no Protestant was ever admitted to see: Our Commission was to Travel for Three Years, and one Master *White* a Protestant, was permitted to Travel under the same conduct, for my sake.

3d. Sigh.

— *And now Lord what obdurate heart is it that will not be concern'd in departing from all his Friends and Relations, and his own native Country to boot! Nay further, in putting Soul and Body into the Hands of Strangers, and contrary minded Men; had'st thou not been the God of the Sea, as well as of the Land, of the whole World, as well as of part of it, this had been a great hazard indeed; but thou art in all places at all times, the Heavens and Earth are thine; there is no way to avoid thy Blessed Presence but by Sin, and no way to embrace it but by Grace.*

In Travel.

In *October* 1636 or 1637, we took Ship at *Dover* and landed at *Callice*,[75] leaving (like young Seamen) some tribute behind us, to gratifie the Fish: There we understood the direct way to *Paris* was obstructed by disposing the *French* Army into their Winter Quarters, and that the Earl of *Bedford* that now is,[76] was Robb'd in his passage thither; we therefore went an Oblique way by Caroch (as they Scandalously call it, *Anglice*, a Wagon) and as I remember, we went to *Oske*, where there was an *English* Nunnery newly erected, and intended to be Consecrated the next day; we had therefore the

[58]

opportunity to see, and speak with the intended Nuns, and view their *Dormitories,* which was never after to be done: From thence we went to *Doway,* where there was an *English* Colledge,[77] and Doctor *Kelleston*[78] the Master thereof, and at present Doctor *Labourne* who was Master *Arundells* Tutor: There we stayed some time, and after to *Cambray* St. *Quintins,* and so to *Paris;* where there was great bemoaning of Duke *Memorancee's*[79] death, the last of that most Ancient, and Noble Family, who for joyning with Mounsieur the Kings Brother, in an Insurrection against the King, was beheaded not long before: We Stayed there about a Month, to fit us for our Winter Quarters in *Orleans.*

Memorancee beheaded.

The most remarkable passage during that time in *Paris* was the beheading of a Noble man's Page, who for some misdemeanour, was condemn'd to Death; and being well beloved of the Pages and Lackeyes, they appeared so numerous, that though there was a Horse-Guard to see the Fact done, they could only secure his person back again to prison: The next day the King commanded 5000 Souldiers to see him Executed, but there appeared near 20000 Pages and Lackeyes Arm'd, that they thought it not fit to take him out of Prison: The next day the King commanded his whole Army in and about *Paris* to see Execution done upon him, and also that the Streets should be Chain'd up, and Proclamation made that whosoever attempted to rescue him, should be proceeded against for his Life; notwithstanding all which, the Pages and Lackeyes appeared so numerous and resolute, that they brought him not out of Prison, but beheaded him there, and hanged his head severed from his body out of the Window; which when the Tumult saw, they dissolv'd, and went their several waies.

Before the Winter was ended, Mr. *Arundell* got a Heat and a Cold at Tennis, which turn'd to a Feaver, and dyed at *Orleans,* whose Death was the end of our Journey, and soon after we return'd to *Paris* again, where we stayed no longer then to prepare for *England.* But before I leave *France,* I must acquaint you with the civility of Doctor *Labourne,* Mr. *Arundell*'s Tutor, whom we left in *Paris;*

Doctor Labourne his civility.

who notwithstanding his Pupils death, used Mr. *White* and my self with all possible kindness as before, and according to his promise made formerly to my Friends in *England,* never striv'd to alter my Religion; but by his good Life and Conversation, which was so exemplary, that all the Musick and Adornments of Churches and Chappels, with all the Religious Houses of Monks, Fryers and Nuns, could not perswade so much.

The passage in our Return to *England* was very dangerous, for when we were within a League of the *English* Shore, our Ship was forc'd back again upon the *French;* but with some help from the *English* Coast, Master *White* and I came safe to *Dover.* One passage I cannot omit to declare, that when I went out of *England,* I resolv'd not to drink between Meals 'till I return'd thither again, which resolution I kept strictly.

A good resolution.

Obj.

And here me thinks I lye so open to the Reader, that he cannot choose but say, What need is there now of Sorrow, Since you are past all dangers beyond Sea, and come safe home; having left the Vice of Intemperance behind you, which was some cause of your Travels to avoid? If you had just cause to Sigh when you were abroad, you have just cause to Rejoyce now you are come home; and 'tis not only Ridiculous but Sinful to Mourn when you should Rejoyce.

Answ.

I confess there is a time for all things, and good Christians should like good Musicians, keep it; but when I consider that it is but the beginning of a great work, that I have not sufficiently repented for doing what I ought not to do; that many Sins are national, and Men are apt to relapse, when they come into the same company with whom they have committed them; and that at best, I have conquer'd but the Beastial part of Man, there is still great cause of Sorrow.

4th Sigh.

— *Lord, give me not only a sense of Sin, but a sorrow for Sin; and since thou hast begun a good Work in me, grant me, I beseech thee, the Grace of thy Holy Spirit to perfect it; and let me not think my self safe when I have Conquer'd the beast within me, but still persist to conquer the man within me, that thou may'st be Glorified by my Salvation, and not by my Damnation.*

From *Dover* (taking *London* in the way, where I met my Uncle *Sandys* Coach) I made all convenient Speed to the *Vine,* where my Father and Mother then were: I need not tell you with what Joy and satisfaction I was received of all sides; and being now taken from the design of Travel, and the Inns of Court, whether to return was thought a degradation; I betook my self to Country affairs, and my Father trusted me much with the managing of his Estate, (which I took some delight in) and alwaies gave me a sufficient allowance for the Imployment I was in; and in those dayes I never knew what want was. After I had been some time there, my Friends furnished me with all accommodations I wanted, as Cloathes, Horses, &c. and Sent me to Count *Arundell* and his Lady, to acknowledge their great kindness to me, and to Complement them as well as I could; the Lady (after she had paid her Funeral Tears to the memory of her only Son) told me that now he was Dead, she would take me for her Son; and I had very good cause to beleive her, but the difference in Religion as well as other respects, would not permit me to be capable of so much Honour; yet till her Death she continued a very great Friend to me. She was a person of the greatest discretion, temper and humility, that think I ever Saw; what other women pretend to do, she did; she brought her Lord out of a debt of 10000 *l.* and several Law Suits; kept a very good Table; bought at least 1000 *l. per ann.* Land, Married five or six Daughters, gave at least 6000 *l.* a peece with them, one with another; and allowed her Lord 500 *l. per ann.* to cast away upon the Philosophers-Stone.

The Dyet I began with in *France,* I continued in great measure in *England,* but not with that strictness as not to drink between meals; but this I added to it, in attempting to keep two Lents, but fail'd in both: The first gave me an Ague, the second a Feaver, and did my constitution much wrong; causing vapours to fly up into

My return to England.

A Wife indeed.

my Head, that I became very sickly. When I was in the Country, I did as a Country-man; kept Hounds and Horses, and us'd much Exercise. When I came to *London*, I did the like; as Dancing, Riding the Great Horse, *&c.* And all this time I kept mostly an abstemious dyet, and Subdued my Body with much Exercise. No Sooner came I of full Age, but my Father Dyed, and left me an Estate; out of which I can modestly say, I could have Spent 800 *l. per ann.* without being in Debt; nor was there any Debt, or Legacy upon it, but only my Mothers Joynture; and I did not owe Ten pounds in all the World. Upon this advance of Fortune, I became more worldly minded than ever; for though I wanted nothing before, I had nothing till then I could call my own: But this continued not long, for after the daies of Mourning were accomplished, I put off my Hounds, put 200 *l.* in my Purse, and came to *London* and kept my Coach: Where Queen Mother[80] preparing for a Masque, and many of the Gentry in Honour to the Queen, intended to be Ante-Masquers; I was also desired to be one, which cost me my Two Hundred Pounds, yet still I lived within Compass of my Estate.

The Queens Masque.

At Court I found my self guilty of three Imperfections, that would hinder my preferment there: A Blushing modesty, a Flexible disposition, and no great diligence; when I returned into the Country, my Mother had found me out a great Heir not far from the *Vine,* and had made a fair passage for me unto her; and though I was then very averse unto Marriage, yet in obedience to her Command I went; and when I came thither, I found my Mistress much of mine own temper, and we presently agreed never to see one another more: I had several other good offers, but none that I approved of, till I heard of my Lady Acheson,[81] but that I shall reserve for another Sigh.

5th Sigh.

— *And now, good God! what a golden opportunity have I lost? Had I with the labouring Bee laid up in Summer against the Winter of mine Age, I had not been now to seek, but misspent Time and Mony can never be recalled.*

THE VINDICATION OF RICHARD ATKYNS

Having used all kind of Sports and delights in *Hantshire* (which was a pleasant Country) and at *London,* I began to dislike my own Estate in the Vale of *Gloucester,* which was so dirty, so enclosed and so far from *London,* that I was not capable of the same Delights there; in addition to which my House was burn'd down, and the place as a Seat spoil'd by the long inhabiting of Tenants, and I had no other habitable House in the Country; All these, together made up an Excuse of leading the same life as before, with this also, that the old Lord *Sandys* had given a Mannor of about 200 *l. per ann.* after a life to his Grand-Neeces, whereof my Mother was one, who with another of my Aunts, gave me their Share, to buy out the rest, and the Life in Posession cost me about 1500 *l.* which was the only Debt I then owed for myself, and I proposed to pay it out of Fines in letting Copy hold Estates, or a Marriage Portion, which was very probable, but afterwards my Uncle *Sandys* (to whom I was much obliged in gratitude) falling into a declining condition by unfortunate Engagements, prevailed upon me to be bound with him in several Sums of Money, for which he gave me sufficient security, but so did not a certain Relation of his, for whom I became also bound: But both of their Debts lying upon me for several years, and the War coming on, the mischief of them could never be recovered. *My first Debt.*

My Engagements and the causes for them.

Whil'st these things were doing, my Mother (being desirous to settle me) had intelligence of a Lady, rendred to be a very great Fortune, who had lately buried her Husband and Father, by whom she had both Inheritance and Joynture, which prov'd to be my Lady *Acheson,* and though, without sight, I had a greater opinion of this than any other I ever heard of, yet the person from whom this report came, was so inconsiderable that I could not believe it, till upon farther enquiry 'twas found to be true: How to have access, was now the business in hand; in order to which the person that first proposed it brought a Scotch Captain to my Mother and me, who declared many particulars as to the Estate, and pretended great acquaintance with the Lady, but could not procure a Visit; he

The first News of the Lady Acheson.

[63]

then brought me acquainted with Captain *Acheson,* a Kinsman of her late Husbands, who used to eat at her Table, but neither could he bring it to pass, but put me in a probable way of seeing her, which was by entertaining in my Service a Scotch man who Waited upon her first Husband, and was well belov'd in the Family, this young man prov'd very honest, and the very best Servant that ever I had, he went frequently thither, and brought me the true News of the House, by which I might the better mannage my design.

After four Months time, I heard she entertained some Gentlemen, whom I conjectur'd might go upon the same Errand I desired, which made me resolve to hazard my Four Months attendance and go my self, from which I was diswaded by my man *Irwing* (for so was he called;) about 2 in the Afternoon I went and took him with me, who finding the the Parlour Door bolted against us, went round about and let me in : The Lady seeing through a peep-hole a Stranger at the Door, went up to dress her self, leaving her two Cousins behind, and supposing it was her God-mother, (who was to Visit her that Afternoon) sent for all the company up to the Dining-Room; whither I went up also, and was put in Countenance by a Gentlewoman of Quality that was there, whom I had seen twice before: Not long after a Countess came thither with her Brother, and the Gentlewoman I had some acquaintance with advised me to take my leave, which I did; and she did me the Honour to go along with me. Upon this success, two dayes after at the Same time of day I went again, and meeting her Woman, she told me her Lady kept Chamber and was not to be seen: I desired her to present my humble Service to her Lady, and tell her I was to wait upon her; and to give me an Answer before I went away, which she did, but staid a long time before she return'd; which time was spent in debate with her Cousins whether I should be admitted or not, at last 'twas concluded I should; I had not been there half an hour, but some visitants came to her again, and I took my leave, which was kindly receiv'd.

My wooing time.

[64]

THE VINDICATION OF RICHARD ATKYNS

The Executors (who would have also been her Guardians, she being under Age) began to suspect me more than all the rest, and laid their heads together to bespatter me, and to spoil my further visiting; and advised her to forbid me the House, which she said she could not well do, because I had behav'd my self civilly, and had not proposed any thing of Love to her, which others had done; and for her to suppose I came to that end, unless I had declared it, would argue too much forwardness in her: This gave them little satisfaction, yet I found it made my Visits less acceptable to her than before; at last a Kinsman of mine hearing by the Ladies Uncle (who was next Heir, and one of the Executors) that I was traduc'd, and assuring him that those things that were spoken of me were to his knowledge false, and desired therefore that he would be an Instrument to bring me to the Lady to vindicate my self; who promised to do it in two daies after, and was as good as his word: When I had vindicated my self, I desired to speak with her in private, and then, and not before, I told her how much I Honoured her, and to confirm the truth of it, I fell down in a sound; these things rais'd thoughts on all hands, and being prepared to Wait on the King in the Scottish Expedition, and my Mistress to go into the Country, upon consideration, and some hopeful words that fell from her, I found Venus more influential upon me than Mars, and staid behind.

At her going out of Town, I took my leave of her, but not of her Kinswomen, because I would not want an excuse to see her the next day, but before Seven in the Morn they were gone, and I after them in a Coach and four Horses, and overtook them within a Mile of *Colbrook;*[82] where we Din'd together, and there I wanted an excuse to carry me further; only, to see them part of their way: Within half a Mile, I observ'd the Cook-maid very un-easie on Horseback, and commanded her to be put into the Coach with me, and then drove up to the other Coach; which when my Mistress saw, and that I was resolv'd to go to *Farringdon;* she allowed me one of her Cousins to bear me company: When we came to *Henly,* the

A Journey into the Country.

best Rooms in the Inn were taken up by Sir *William Sandys* my Uncle, who for my sake, quitted the very best Room to her: The next Morn, after I had helped my Mistress into the Coach, I went in my self, and sat by her, and handed her Cousins in after me; and with much adoe, prevailed that her Servants and Baggage might be carryed in mine: From thence we went to *Farringdon,* where my Mistress had a House in the Town, I staid a Week, and was all day at my Mistresses House, and so to *London* again. But the Executors hearing of this, writ many discontented Letters to her, which were still rubs in my way, but I made Weekly visits in the Summer, whilst she was in the Country; and in the Winter she came up to Town, where we were upon a Treaty of Marriage; but the Executors still zealous to break off the Match, enquired, and found out that I was engaged for Money as before, and prest that so home, that they were very likely to accomplish their design.

6th Sigh. — Had I read the Scripture with that zeal, and frequency as Playes and Romances, I must needs have discovered the misery of Suretyship; for saith Solomon, *why should thy Bed be taken away from under thee?* and King David *adviseth, to honour our Creator in the dayes of our Youth: How sad a condition are they in then that serve the Devil with their affection and strength, and God with their decrepit old Age; and how much sadder those who serve him not at all, who do not only with the Fool, say in their Heart there is no God, but more lively express it in their Words and Worship?*

By this time Love began to play the advocate for me with my Mistress, and all disswasions prov'd but as whet-stones to sharpen the appetite of desire, the thing now to be done was to settle my Estate, and make her a convenient Joynture out of it, which being left to the Executors, went on very slowly; and many Exceptions were taken to my Estate, though my Mother to help me in this great Work, was willing to Surrender her Joynture, and receive recompense out of a Lease holden of the Dean and Chapter of *Gloucester.*

Whilst these things were agitating, the Queen had another Masque, wherein I had a part, and obtained the Honour of my

Mistress to see it, and because she should not be throng'd, I invited her to a Supper at Court, and had a promise from the Lord *Chamberlain* that she should be well Seated; but before we were called up, the House was so filled with ordinary People, that several Countesses of the better sort could not be admitted to see it, which was the cause it was represented again, and if I had not refused to go in without my Mistress, she had not been admitted neither; In the throng I lost a Diamond Hatband of Three Hundred Pounds price,[83] which being carryed to my Lord *Chamberlain* as his Hatband I had again, but my Mistress got so great a distemper that the next day she fell Sick of a Quincy,[84] sent for me and put her self into my hands, and would have no Physician but of my providing; She also settled her Estate upon me as well as she could (being under Age) and sent for my own Uncle to do it. Before she was fully recovered, the Queen commanded the Masque to be represented again, which I hearing of, gave my Habits to a Dancing Master and Five Pounds to perform my Part; and sent to the Lord *Chamberlain* to desire him to excuse my so doing: The Queen exprest a gracious resentment of the sorrow I was under, but withal desired I might perform my Part my self; saying, she very well liked that entry, which coming to my Mistress Ear, she also desired me to do it. By this time I was Pursy[85] and Feaverish by lying in the House, and the Dancing Master proud of my Part, would not let me have my Habit again but cut it less for his own Body, which with a Trowse I bought to wear under it; that was to big for me, caused me to role it up and put it into my Breeches, but upon rising of Capers it broke my Buttons before, and started out of my Codpiss, which was the occasion of a great deal of Mirth and Laughter, the greatest part not knowing what they Laught at; amongst which the *Venetian* Embassadors Lady bearing her part, was severely chastised by her Husband, which she upon a visitation to then Master of the Ceremonies Lady declared, asking also, whether all the *English* Ladies that Laught were so serv'd: She answer'd, 'twas not the fashion of *England* to do so, and she was very sorry 'twas the fashion of

Another Masque of the Queen.

Love Exprest.

A foul fact on a fair Lady.

Venice, and that the Gentleman for whom she suffered, was then in the House; but because she knew it to be the mode of *Venice* not to entertain a person of her Quality where a strange Gentleman was, she commanded her Daughters to entertain me in another Room; to which she replyed that she had Suffered often before, but never for a man whose Face she never saw: Upon which the Lady *Finnett*[86] caused her Daughters to make some excuse to bring me through the Room, and so back again; but I never knew the reason of it till afterwards, with which she seem'd very well satisfied.

By this time my Mistress was pretty well recover'd, the Writings were agreed upon and drawn, and she made what settlement she pleas'd upon my Estate, without one penny portion in mony; and her Estate (all but *Farringdon*) in her Uncles hands: And because we would have no more rubs, we agreed to Marry presently, which was in Lent, about the 20 of *March* 1640. when She was between 19 and 20 years of Age, and my Self between 25 and 26. She made choice of her Uncle to give her in Marriage , which he refusing , Sir *John Finnett* Master of the Ceremonies, did it most willingly. And now pretenders being laid aside on the one hand, and adversaries conquer'd on the other; I took my self to be in an unshaken Kingdom and exprest it in the Poesy of my Wedding Ring, which was much disliked; and I believe no persons were ever married together that loved better.

7th Sigh. — *But Lord there is no compleat happiness in this World! the end of one misery is the beginning of another; when we think our selves most firmly establish'd, we are set but in slippery places, which we shall never discover till we come into thy Sanctuary; for man being in Honour having not understanding, is like the Beasts that perish.*

Upon this advance of Fortune, which the World esteem'd no less than 20000 *l.* as also, that I had Married a Beautiful and Discreet Woman: My credit grew so high that I could borrow but too much Money; my Uncle Sandys therefore to preserve the *Vine,* earnestly importun'd me to joyn with him in a Judgment of 3500 *l.* for payment of 2500 *l.* without which he could not have the Money (which was the first binding Engagement I ever entred into) and a pretended Friend advis'd me to comprize the several debts; I was bound in for one Mr. *W. S.* in one Sum, to avoid brokeage and preserve my credit, and he would lend me the Mony, which I consented to; but when I came to give him security for it, he made himself a Lease of 99 Years of a Mannor, and a Statute of 6000 *l.* to perform Covenants; and perswaded me to give a Statute of so great a Sum; for that if I wanted 1000 *l.* more, he would lend it me upon that Security: And here I found the want of an Inns of Court breeding, not knowing the utmost danger of a Judgment, or Statute: However, I had such promises and protestations from both of them I was engaged with, that they would pay these Debts within half a Year, that I must have been an Infidel not to have believ'd them; and this was the worst thing that ever I did in my life (as to worldly Affairs) and hath been the cause of all my future Misery.

My first binding Engagements.

These Debts lying heavy upon me, and the War coming on, would not suffer me to pay my own Debts, (which were then reduced to 1000 *l.*) except I paid theirs first, which when I discovered, afflicted me very much; and then, and not till then, I lookt into my Wives Estate, which I found little or nothing serviceable to me, for her Uncle (who was the most passionate and inconsiderable man that ever I knew trusted with Such an Estate) had possest himself of the whole Estate in the *Strand,* both real, and personal, (except some House-hold-stuff) by virtue of a Lease of 21 Years, whereby he was Trustee; and by alone proving the *Will,* whereby he was become sole Executor, and it was incumbred with Debts and Legacies also; the place most free, and whereof she was possest, were Lands in *Farringdon,* part whereof was litigious by the claim of Sir *Robert Pye,*[87] another of her Uncles had an Annuity thereout, and the whole lyable to a Charitable use: 'Tis true there were some Houses made over upon her first Marriage with Sir *Patrick Acheson;* part whereof were destroyed which ought to have been

A great cry and little weoll.

[69]

received by her quietly, but the receipt of the remaining Houses were also interrupted by her Uncle: There were also Several other Rents she ought to have had; as 600 *l. per ann.* Joynture out of *Ireland,* part whereof was Mortgaged by Sir *Patrick Acheson* her first Husband, after a Joynture made; and the other part possest by his Creditors, who obtained the Lord *Straffords*[88] Order for possessing the said Lands till their Debts were paid; albeit they were made in Joynture before, there was also a Rent charge out of *Micheldever,* a Patent for Printing, a Salt Work, and several Adventures, in which her Father had a share; but all possest by her Uncle: This was the condition I found her Estate in. I went therefore to the best Councel I could get, to advise upon the said *Will,* who found it so intricate (as it was prov'd with Codicels) that they advised me to agree if possible I could; I did then offer her Uncle more than was his due by the *Will* prov'd, and to give security to perform it; and several agreements were made to that purpose, but all in vain, for he stood to none of them; which spent a great deal of time: at last I was advised to put in a Bill in Equity against him, and the rest of the Executors; which with much adoe, my Wife consented to; but this was to little purpose, for before I could bring it to a hearing, the War began, and I returned into the Country with these Debts upon me; and here might I cry with King *David.*

8th Sigh. — *Save me O God, for the Waters are come in even unto my Soul: I stick fast in the deep Mire where no ground is; I am come into deep Waters so that the Floods run over me: but thou O Lord art my Defender, and the lifter up of my Head; O deliver my Soul! and save me for thy mercy sake; so shall mine Enemies be confounded, they shall be turned back, and put to shame suddainly.*

As no Cities nor Counties were free from preparations for War as their affections inclin'd them, so the parts about *Gloucester*[89] (to which I retired,) happened to be most Unanimous for the Parliament, which was contrary to my Judgment: For I am perswaded, that none that heard the Lord *Straffords* Tryal,[90] and weigh'd the

THE VINDICATION OF RICHARD ATKYNS

Concessions of the King in Parliament, could conscientiously be against him; but 'twas now too late to remove from thence; for Fears, and jealousies, had so generally possest the Kingdom, that a man could hardly travel through any Market Town, but he should be ask'd whether he were for the King, or Parliament. In this nick of time my Servant *Erwing*, (who presently after my Marriage, betook himself to the *Gen d'Arms* in *France*) hearing of the War in *England* came over and proffered his Service to me again; whom I received as before: And being well known to his fidelity, I sent him to *London* to his Brethren the *Scots,* to give me the best intelligence he could; who did it most truly, and prophetically; and as an Argument of his affection to me, refus'd a Lieutenants Place of Horse on the Parliaments Side, to continue my Servant. Him I imployed to train up my Horse, and make them bold; under one *Forbes* his Countryman, who was then Governour of *Gloucester;* and soon after the Battail of *Edghill,*[91] I waited upon the Lord *Chandos*[92] to *Oxon,* not intending at that time, to stay any longer than to present my self to the King, and to assure him of my Duty and Affections: But whilst I staid there, I received intelligence, that my being at *Oxon* was publiquely known at *Gloucester;* so that I could not return in safety; but sent for the men and Horses I left behind to come thither after me, and when the Lord *Chandos* had accepted of a Commission to raise a Regiment of Horse, and Mustred his own Troop, he gave me a Commission for a Troop under him; which I rais'd with such success, that within one Month, I Mustred 60 men besides Officers; and almost all of them well Arm'd; Master *Dutton* giving me 30 Steel Backs, Breasts and Head Peeces, and two Men and Horses completely Arm'd: And this was done upon Duty, without any advance of Mony, or Quarters Assigned; wherein every 4th or 5th man was lost.

But as some were lost, others were added; For one *Powell* a Cornet of the Parliaments, with two Troopers, (all very well Horst and Arm'd,) came into my Troop at *Oxford:* I carryed him to the King, and beg'd his pardon, which the King graciously granted; and in

A faithful Servant worth gold.

My raising a Troop.

token thereof, gave him his Hand to Kiss; but ask'd him for his Commission, which when he saw; he said, he never saw any of them before; desired to keep it, and put it up in his Pocket, and gave him good councel, with very great expressions of his Grace and Favour to me.

My Troop I paid twice out of mine own Purse and about a fortnight after, at the Seidge of *Bristoll,* I mustred 80 men besides Officers; whereof 20 of them Gentlemen that bore Armes: (Here the Swearing Captains, put the name of the Praying Captain upon me, having seen me sometimes upon my Knees.) The Lord *Chandos* afterwards (though I had the Honour to be allyed to him) us'd my Troop with that hardship, that the Gentlemen unanimously desired me to go into another Regiment; which his Lordship understanding, thought to affix me to his by a councel of War; but failing therein, I was admitted into Prince *Maurice's* Regiment,[93] which was accounted the most Active Regiment in the Army, and most commonly plac'd in the our Quarters; which gave me more proficiency as a Souldier, in half a Years time, than generally in the Low Countries in 4 or 5 years; for there did hardly one Week pass in the Summer half year, in which there was not a Battail, or Skirmish Fought, or beating up of Quarters; which indeed lasted the whole Year, insomuch as for three Weeks at most, I commanded the Forlorn-hope thrice.

Forlorn-hopes.

The first, was at *Little Deane,* in the Forrest of *Deane* under the Conduct of the then Lord *Grandison,*[94] against Sir *William Wallers* Army; in which was a remarkeable accident: For no Sooner had I received the word of Command, but my charging Horse fell a trembling and quaking that he could not be kept upon his Legs; so that I must loose my Honour by an Excuse, or borrow another Horse presently; which with much adoe I did of the Lord *Chandos* his Gentleman of the Horse, leaving twice as much as he was worth with him; The Charge was seemingly as desperate as any I was ever in; it being to beat the Enemy from a Wall which was a Strong Breast Work, with a Gate in the middle; possest by above 200 Musquetteers, besides Horse: We were to charge down a Steep plain Hill, of above 12 score Yards in length; as good a Mark as they could wish: Our party consisting of between Two and Three Hundred Horse, not a man of them would follow us: So the Officers, about 10 or 12 of us, agreed to Gallop down in as good Order as we could, and make a desperate Charge upon them; the Enemy Seeing our resolutions, never Fired at us at all, but run away; and we (like young Souldiers) after them, doing Execution upon them; but one Captain *Hanmer* being better Horst than my self, in persuite, fell upon their Ambuscade and was killed Horse and Man: I had onely time enough to turn my Horse and run for my life. This party of ours, that would not be drawn on at first, by this time, seeing our success; came into the Town after us, and stop'd our retreat; and finding that we were persu'd by the Enemy, the Horse in the Front, fell back upon the Rear, and they were so wedg'd together, that they routed themselves, so as there was no passage for a long time: All this while the Enemy were upon me, cutting my Coat upon my Armour in Several places, and discharging Pistolls as they got up to me, being the outermost man, which Major *Cheldon*[95] declared to my very great advantage: But when they persu'd us to the Town, Major *Leighton* had made good a Stone House, and so prepared for them with Musquetteers; that one Volley of Shot made them retreat: They were so neer me, that a Musquet Bullet from one of our own men took off one of the Bars of my Cap I charged with, and went through my Hair, and did me no hurt: But this was onely a Forlorn party of their Army to face us, whilst the rest of their Army March'd to *Gloucester.* At my return to the Rendezvous, 'twas debated at a Councel of War, whether to quarter in the Field all night, or to March to *Tewxbury;* from whence we had drawn too many men: The first was resolv'd upon; but the Enemy finding their advantage, put their Garrison Souldiers into Boats (who were fresh) and their wearied Souldiers in their rooms, and surpriz'd our Garrison of *Tewxbury* that night; of which we had intelligence by Nine the next Morning.

A great deliverance.

The surprise at Tewxbury.

RICHARD ATKYNS ESQUIRE

A skirmish at Ripple-Feild.

The next Forlorn-hope I was commanded upon, was with Major *Cheldon*, about 4 or 5 Daies after at *Ripple-Feild*,[96] where we did good Execution upon the Enemy; but the Lord *Grandison* (as I take it) received at that time a hurt, of which he dyed. We were then Commanded to March towards *Oxon*, to the relief of *Reading*, beseidged by the Earl of *Essex*,[97] with the Kings whole Army that Quartered thereabout: Where I had the Honour to Command the Forlorn hope again; 8 or 10 Daies after, the general Rendezvous was by *Wallingford;* and the Forlorn-hope consisting of about 160 Horse, was sent out between 8 and 9 of the clock in the Morning; we March'd to *Nettle-bed,* and so to *Cawsam-house*;[98] through as bad way for Horse to March as ever I saw: For the Way was so thick of Woods, and Furzes, that in two Miles we could not draw up 8 in Front, so as a small party of Horse might easily have retarded our march, and killed several of us. Our security might be the illness of the way; for none that knew not of our March before, would ever expect Horse to March in that place: We saw not

Cawsam fight.

one Scout or Armed man, till we approach'd *Cawsam* Bridge, and there we found their Army prepared to entertain the Kings; My Station was (not directly, but Obliquely) between the River and a large Barn, within Musquet shot of both; they sent no party out to Fight us; but within half an hour, the Kings Army appear'd upon a Hill, about a Mile off (him-self being in Person there.) The Cannon play'd upon us, but did us no harm; we Kill'd some and took others Prisoners, they mistaking us for their own Party: Between 12 and 1 of the Clock, the King sent down several Regiments to Storm the Barn; without the taking of which, we could not have access to the Body of their Army, which lay mostly between the Barn and the Bridge: And as the King alwayes adventur'd Gold against Silver at the best, so now he adventur'd as Gallant men as ever drew Sword, against mud Walls; for the Barn was as good a Bulwark as Art could invent. 'Twould grieve ones Heart, to see men drop like ripe Fruit in a strong Wind, and never see their Enemy; for they had made loop-holes through the Walls, that they had the full

[74]

Bodies of the Assaylants for their Mark, as they came down a plain Field; but the Assaylants saw nothing to shoot at but mud Walls, and must hit them in the Eye, or loose their Shot. Upon this disadvantage I need not tell you what men we lost; about Three of the Clock, my Party was relieved for half an Hour, and then the Party that relieved us were drawn off again.

Soon after, the Kings Army March'd off (having releiv'd the Town with Ammunition) and my Party was left as before, without hope of releif, and the Sun going down; all which daunted my men so much, that I could hardly make a Front of 6 men; and indeed the Danger was not small, for Two or Three Hundred Musquetteers had lin'd the Hedge by this time, within half Musquet Shot of us and began to play upon us on the one side, a Regiment of Foot and Cannon about Musquet Shot fronted us, and a strong Party of Horse on the other side; so as I had much adoe to keep them from running, having a Leiutenant as fearful as any; which to prevent, I was forced to cut some of them, and threaten my Leiutenant; with which we stuck together, more like a flock of Sheep, than a Party of Horse; until Prince *Rupert*[99] sent his Commands by Collonel *Legge*,[100] that I should March off, and make my retreat as well as I could; which by the advice of that Noble Person, I did with such success, that I lost not one man. About half an Hour within Night as we March'd in a broad Rode, my Scouts discovered the Enemy, and came very merrily back; at which my Party took such a fright, that though I desired them with all importunity (for Command was now laid a Sleep) that but six of them all with fixt Pistolls would go along with me, and I could get no more than two: By this time Draggoones on both sides the Rode were pelting at us, but at a great distance; and a party of Horse persu'd us in the Rear; but when they came in a convenient distance of us, we three discharged at them, and they ran away with so great a noise, that my Party suspecting they had come on, ran too, but the contrary way; we were afterwards troubled with them no more: And about a quarter of a Mile further, Prince *Rupert* had laid so strong an Am-

A deliverance.

Clinias *and* Dametas.[101]

buscado for them, that if their whole Army had persued us, I'm confident he had scatterred them: Upon this service I was not a quarter of an Hour in fourteen Hours off my Horses back, and Prince *Rupert* declared he would never put me upon so hard duty again.

9th Sigh.

— *Lord thou hast been my God, even from my Mothers Womb! Thou hast delivered me from my cruel Enemies, in the time of greatest danger, thou art a God Mighty in Battail, the Lord of Hosts is thy Name: Flies and Lice receiving Commission from thee, can do wonders! and the greatest Potentates without it, shall fly before their most despised Enemies.*

The Advance into the West.

Not long after, Prince *Maurice* his Regiment with others, March'd into the *West*, to assist the then Marquess of *Hertford*[102] in raising an Army in those Parts (who was made General of the *West*) where Sir *William Waller*[103] had been with his Army before, to raise Forces for the Parliament. Prince *Maurice* had such an intire affection to the King, that (not regarding his own Dignity) he took a Commission under the Marquess, rather than the Kings cause should fail. The Lord *Carnarvan*[104] was General of the Horse, and Sir *James Hamilton* Major General: We did not much in our March, but raise Men and Armes, till we came to *Crock-horne*,[105] where (as I take it) we met the Lord *Hopton*[106] with his Forces from *Cornwall* (having clear'd those parts) which were most upon Foot: This addition made us up a pretty marching Army. When we came together, we were quickly upon action; but the *Cornish* Foot could not well brook our Horse (especially, when we were drawn up upon Corn) but they would many times let fly at us: These were the very best Foot that ever I saw, for Marching and Fighting; but So Mutinous withal, that nothing but an Alarm could keep them from falling foul upon their Officers. The first thing that we attempted (as I remember) was the taking of *Taunton Dean*,[107] which stood out at first, but when we were prepared to Storm it, they yeilded: I having then a Commission for raising a Regiment of Dragoons; my self, and Some of my Officers went to seek for Arms; of which we

Taunton Dean taken.

found many, and observing a hole in an Elder hedge, I put in my hand and took out a Bag of Mony; which if our Foot had espied (who were also upon the Search) they had certainly taken me for the Enemy, and deprived me of both it and Life.

After the Garrison settled in *Taunton Dean,* we March'd towards the Enemy, who were then Quartered in, and about the *Bath;* we had 2 very hard daies March thither: But before we came to *Glastenbury,* we had intelligence of some Forces newly rais'd, Marching towards their Body; whom a Commanded Party of ours persu'd, and did some Execution upon: That day came Sir *Horatio Carew* in to us. When we came to *Wells,* intelligence was given us, that Sir *William Wallers* Army was drawn out on that side of *Bath;* we March'd toward them as far as *Chuton,* which I suppose, is about half way; the Sun was then about an hour high, and many of our Horse and Foot tyr'd with our March; so the Foot had Orders to Quarter at *Wells,* the head Quarters, and the Horse thereabouts: The Quarter-masters were sent to take up Quarters accordingly; and the Lord *Carnarvon* with his Regiment of Horse, went to give their whole Army an Alarm; but came so near them, that for hast, they sent out a fresh Regiment of Horse, and another of Dragoons to fight him; his Lordships Regiment being much wasted, and his Horses tyred with the long March, were forced to retreat; and the Enemy had the persuit of them to *Chuton,*[108] where Prince *Maurice* was hurt, and taken Prisoner. We were then a Mile on our way towards our Quarters, when Collonell *Brutus Bucke* acquainted our Regiment with this unwelcome news; which I heard first, having the Honour to Command the Rear Division of the Regiment. My Leiutenant Collonel, my Major, and the rest of the Officers, advised what to do in this case; and the result was, that Prince *Maurice* having himself Commanded his Regiment to their Quarters, they were Subject to a Councel of War, if they should disobey Command; to which I Answer'd (being eldest Captain) that I was but a Young Souldier, and if they would give me leave, I would draw off my Division and run the hazard of a Councel of War; they

Chuton *Fight.*

Prince Maurice *taken Prisoner.*

RICHARD ATKYNS ESQUIRE

told me, they might as well go themselves, as give me leave to go; but if I would adventure, they would not oppose it, but defend me the best they could.

I drew off my Division with all possible Speed, and put them in order, which were not above 100 men; and before we had March'd twelve score Yards, we met the Lord *Carnarvan's* Regiment scattered, and running so terribly, that I could hardly keep them from disordering my men (though in a large Champaign)[109] at last I met his Lordship with his Horse well nigh spent, who told me I was the happiest sight he ever saw in his life: I told him I was no less glad to see his Lordship; for as yet I had no Command for what I had done, and now I hoped he would give me Command publiquely, to preserve me from the censures of a Counsel, which he did. The Enemy seeing a Party make towards them, left their persuit, and drew up at *Chuton,* and the Lord *Carnarvan,* the Lord *Arundel* of *Wardour,* with my self, March'd in the head of my Party; this was about half an hour before Sun set; and when we came within 20 score of the Enemy, we found about 200 Dragoons half Musquet Shot before a Regiment of Horse of theirs in two Divisions, both in order to receive us. At this punctilio of time, from as Clear a Sun shine day as could be seen, there fell such a suddain Mist, that we could not see Ten Yards off, but we still March'd on; the Dragoons amaz'd with the Mist, and hearing our Horse came on; gave us a Volley of Shot out of Distance, and disordered not one man of us, and before we came up to them, they took Horse and away they Run, and the Mist immediately vanished. We had then the less Work to do, but still we had enough; for there were 6 Troops of Horse in 2 Divisions, and about Three, or Four Hundred Dragoons more, that had lined the Hedges on both Sides of their Horse; when we came within 6 Score of them, we mended our pace, and fell into their left Division, Routing and Killing several of them.

A Mist from Heaven.

The Dragoons on both sides, seeing us so mixt with their men that they could not Fire at us, but they might kill their own men as well as ours; took Horse and away they run also. In this Charge,

THE VINDICATION OF RICHARD ATKYNS

I gave one Captain *Kitely* quarter twice, and at laſt he was killed: The Lord *Arundel* also, took a Dragoons Colours, as if it were hereditary to the Family so to do; but all of us overran the Prince, being Prisoner in that Party; for he was on Foot, and had a hurt upon his Head, and I suppose not known to be the Prince. My Groom coming after us, eſpied the Prince, and all being in confusion, he alighted from his Horse, and gave him to the Prince, which carryed him off: And though this was very great success, yet we were in as great danger as ever; for now we were in disorder and had ſpent our Shot, and had not time to charge again; and my Leiutenant and Cornet, with above half the Party, followed the Chase of those that ran, within half a mile of their Army; that when I came to Rally, I found I had not 30 men; we had then Three Fresh Troops to Charge, which were in our Rear; but by reason of their Marching through a Wainshard, before they could be put in order: I told those of my Party, that if we did not put a good face upon it, and charge them presently, before they were in Order, we were all dead Men or Prisoners; which they apprehending, we Charged them; and they made as it were a Lane for us, being as willing to be gone as we our selves. In this charge there was but one of my Troop Killed, and 8 hurt. For the wounded men of my Troop, and also of my Division I received 20 *s.* a man of Sir *Robert Long*, then Treasurer of the Army; which was all the Money I ever receiv'd for my Self, or Troops, during the War.

The Prince rescued by an inconsiderable Person.

A great deliverance.

When I came to *Wells*, the Head Quarters; I was so Weary that I did not my Duty to the Prince that night, but laid me down where I could get Quarters; I was much unsatisfied for the loss of my Leiutenant and Colours, of which I had then no account; and laid all the Guards to give me News of them, if they escaped. Early in the morning, Mr. *Holmes* my Cornet brought my Colours to me, which pleas'd me very well; but with this allay, that my Leiutenant *Thomas Sandys*, my neer Kinsman, was taken Prisoner, and one more Gentleman of my Troop with him; and that he with some few Troopers, took such leaps that the Enemy could not follow them, else they had been taken also. The next morning I waited upon Prince *Maurice,* and presented him with a Case of Pistols, which my Uncle *Sandys* brought newly out of *France;* the neatest that ever I saw, which he then wanted; but as yet he knew not the man that mounted him, nor whose Horse it was: When I Saw the Horse I knew him, and the man that rid him that day; who was the Groom aforesaid: The Prince told me he would not part with the Horse, till he saw the man that Horst him, if he were alive, and Commanded me to send him to him; which I did that day, and when he came to the Prince, he knew him, and gave him 10 broad Peices,[110] and told him withal, that he should have any preferment he was capable of. This graceless fellow went from my Troop, and took two Troopers with him, none of which ever return'd again: About 15 Years after I saw him begging in the Streets of *London,* with a Muffler before his Face, and spake inwardly, as if he had been eaten up with the Foul Disease. That day I went to my Quarters at *Glastenbury,* where there was a handsome Case of a House, but totally plundred, and neither Bread nor Bear in it; but only part of a Chedder Cheese, which looking blew, I found my Foot-boy giving to my Greyhounds, and reproving him for it; he cry'd, saying there was nothing else to give them: For this Chedder Cheese, I was arrested in London, which cost me 100 *l.*

The Sunday following I desired Doctor *Cole,* Prince *Maurices* Chaplain, to give me and my Troop the Sacrament, which he was willing to do. That morning about 6 my Major *Thomas Cheldon* called upon me, to Dine with Prince *Maurice;* who had invited his Officers to a Buck: I told him what I was to do, *Hang't, hang't Bully, said he merrily thou mayst receive the Sacrament at any time, but thou cannst not Eat Venison at any time.* But that Reverend Doctor gave me, and all my Troop the Communion that day; and I hope 'tis no vanity to say, this small Party probably preserv'd the whole Army; for had these two Regiments carryed off Prince *Maurice* his person, and cut off the Lord *Carnarvon,* and his Regiment, and given an Alarm to our surbated Foot, and tyr'd Horse, being divided, I know not what had become of us. There was one, intituled to this Action (which was usual in the Kings Army) whom I am sure never Struck Stroke in it, (but was within distance to have fallen into the Rear of the last Three Troops, and to have help't us if he would) but there being no good under-standing between Prince *Maurice* and Marquess *Hertford's*[III] Regiments, their Chaplains and Secretaries, whose Places were not to Fight, or to be in the Field; gave Seldome true intelligence, but rather as their affections lead them, supposing that if such things were done, they must be done by such and such Persons.

— Lord thou art the defender of those that put their trust in thee! it is God that girdeth me with Strength of War; thou teachest mine hands to fight, and makest mine Enemies to turn their backs upon me: Thou that hast caused Hezekiah's *Sun to go back, and the Red Sea to be dry, to separate between the* Israelites *and the* Egyptians, *hast sent mist to be a protection to us, and a terrour to our Enemies; not unto us O Lord, not unto us, but unto thy name be the Glory.*

_{10th Sigh.}

After we had refresh'd our selves about a Week in Quarters, we began to seek out the Enemy, who were not far off; for four or five dayes, we skirmish'd by Parties every day, and kept our Body close together expecting Battail daily. Each Army consisting of about 6000 Horse and Foot, but theirs thought to be most; our head Quarters were *Marshfield*,[112] theirs *Bath,* within five Miles of each other: Very early in the morning we sent out a Party of Horse, about 300, commanded by a Major, who did it so ill, that encouraged the Enemies Forlorn-hope to advance so far, as to give a strong Alarm to our whole Army; and we were forced to draw out in hast: The Ground we stood in, was like a streight Horn, about six score Yards over at the end towards *Marshfield,* and twenty Score over at the end towards their Army; on both sides enclosed with a Hedge, and Woods without that. They stood upon a high Hill which commanded us, that opened to a large Down, from whence they could discover our Motions, but we could not theirs; both Bodies within

The Battail of Tog-Hill.

two Miles of each other. For four or five Hours, we sent Parties out of each Body to Skirmish, wherein I think we had the better; but about 3 of the Clock they (seeing their advantage) sent down a Strong Party of Horse, Commanded by Collonel *Burrell*,[113] Major *Vantruske* and others; not less than 300, and Five or Six Hundred Dragoons on both sides of the Hedges, to make way for their advance, and to make good their retreat. And this was the boldest thing that I ever saw the Enemy do; for a Party of less than 1000 to charge an Army of 6000 Horse, Foot, and Cannon, in their own ground, at least a Mile and an half from their Body.

A bold Action by the Enemy.

Our Horse being placed before our Foot and our Cannon, were commanded off Troop by Troop; and being within half Musket Shot of the Hedges lin'd on both Sides by their Dragoons; Several Horses were kill'd, and some of our men; their Musquets playing very hard upon our Horse, made us retreat so disorderly, that they fell foul upon our Foot; and indeed there was not room enough for us to retreat in order, unless we had gone upon the very mouths of their Musquets: I suppose the *Stratagem* was to draw on their Party of Horse upon our Foot and Cannon, the better to rout them, and then our Horse to fall in upon them to do Execution; for the Dragoons making their way by Pioneers, were not discovered till they Shot. Our Commanders Seeing the Army in such disorder, and the Enemies Horse Marching near us; commanded the then Marquess *Hertfords* Life-guard of Horse to Charge them, who never charged before; which was then Commanded by that Honourable and Loyal Person the Lord *Arundell* of *Trarice*:[114] I seeing all like to have been lost, unless a suddain Check were given to this Party of Horse; desired him to give me leave to charge with him, with these words, *That we would answer for one another that day:* We charged together, and both of us fell upon the Commander in Chief, and hurt him so, that heel'd and he reel'd off, and the Party with him; there were several others hurt and kill'd on each side, of the Marquesses Life-guard Mr. *Lee* and Mr. *Barker,* Gentlemen of Quality.

A timely Check given to the Enemy.

THE VINDICATION OF RICHARD ATKYNS

'Twas now no time to draw out a Party of Commanded men, but the Lord *Carnarvan* (according to his usual course) drew up his Regiment as soon as possible, and persu'd them almost to their Body, and kill'd and took several Prisoners; in which Charge (or soon after) he had a Shot in the Leg, that disabled him for further Service at that time. The Enemy to encourage us to prosecute this Success, gave all the Symptoms of a Flying Army; as blowing up of Powder, Horse and Foot running distractedly upon the edge of the Hill, for we could see no further: These signes made Sir *Robert Welsh*[115] importunately desire the Prince to have a Party to follow the Chase, which he gave him the Command of, and me of the reserve; but when he came up the Hill, and saw in what order they lay, he soon quit his imployment there; and desired he might have my Command and I his, which was ordered accordingly. As I went up the Hill, which was very steep and hollow, I met several dead and wounded Officers brought off; besides several running away, that I had much ado to get up by them. When I came to the top of the Hill, I saw Sir *Bevill Grinvills*[116] stand of Pikes, which certainly preserv'd our Army from a Total rout, with the loss of His most pretious life; They stood as upon the Eaves of an House for steepness, but as unmovable as a Rock; on which side of this stand of Pikes our Horse were, I could not discover; for the Air was so darkned by the smoak of the Powder, that for a quarter of an Hour together (I dare say) there was no light seen, but what the fire of the Volleys of Shot gave; and 'twas the greatest Storm that ever I saw, in which though I knew not whether to go, nor what to do, my Horse had two or three Musquet Bullets in him presently, which made him tremble under me at the rate, that I could hardly with Spurs keep him from lying down; but he did me the Service to carry me off to a lead Horse, and then dyed: By that time I came up to the Hill again, the heat of the Battail was over, and the Sun set, but still pelting at one another half Musquet Shot off: The Enemy had a huge advantage of ground upon our men, for their Foot were in a large Sheep-cot, which had a stone wall about it as good a defence

Sir Bevill Grinvill *Slain.*

[83]

against any thing but Cannon as could be, and ours upon the Edge of the Hill, so steep that they could Hardly draw up; 'tis true there were shelves near the Place like *Romish* Works, where we quartered that night, but so shallow that my Horse had a Bullet in his neck: We pelted at one another till half an hour before day, and then we heard not any noise, but saw light matches upon the wall, which our Commanders observing, sent one to discover whether they had quit the Feild or not, who brought news that they were gone.

There were killed of Officers that day, Sir *Bevill Grinvill*, Major *Lower*, Lieutenant Collonel *Wall*, &c. Hurt, the Lord *Carnarvan*, Collonel *Bennett*, &c. and several other Officers taken Prisoners; and more than all these run away to *Oxon*, to carry tydings of our defeat before it was. At the Councel of War that night, were Prince *Maurice*, the Lord *Hopton*, Sir *James Hamilton*, Major *Cheldon*, and some others; the result of that Councel was, that if the Enemy fell upon us, every man to shift for himself: In order to which, the Cannon were drawn off; so that this Battail was so hard fought on both sides, that they forsook the Feild first, and we had leave so to do. The next morning was very clear and about half an hour after Sun rising, we Rendezvous'd our Horse and Foot upon *Togge-Hill*, between the Hill where we Quartered all night, and *Marshfeild;* Major *Cheldon* and my Self, went towards the Lord *Hopton*, who was then viewing the Prisoners taken, some of which were carried upon a Cart wherein was our Ammunition; and (as I heard) had Match to light their Tobacco; the Major desired me to go back to the Regiment, whilst he receiv'd orders of his Lordship: I had no sooner turn'd my Horse, and was gone 3 Horse lengths from him, but the Ammunition was blown up, and the Prisoners in the Cart with it; together with the Lord *Hopton*, Major *Cheldon*, and Cornet *Washnage*, who were near the Cart on Horse back, and several others: It made a very great noise, and darkened the Air for a time, and the Hurt men made lamentable Screeches. As soon as the Air was clear, I went to see what the matter was; there I found his Lorship miserably Burnt, is Horse sing'd like pare'd leather, and *Thomas*

The Ammunition Blown up.

Cheldon (that was 2 Horse lengths further from the blast) complaining that the fire was got within his Breeches, which I tore off as soon as I could, and from as long a Flaxen head of Hair as ever I saw, in the twinckling of an eye, his head was like a *Black-Moor;* his Horse was hurt, and run away like mad, so that I put him upon my Horse, and got two Troopers to hold him up on both sides, and bring him to the head Quarters, whilst I March'd after with the Regiment.

The Enemy (having intelligence of this disaster, and also a recruit that night of Fresh men) gave us no time to repair our losses; but March'd up to our head Quarters, before we could bury our dead, or make provision to secure our wounded; but *Washnage* dyed and gave me his chargeing Horse, which I much wanted; the rest (which were many) were provided for, either in the Marquesses Coach, or Litters made with Boards, except Maj. *Tho. Cheldon,* who was left to the mercy of the Enemy; which he perceiving made shift to get to the Rendezvous, and when he found there was nothing but a Cart provided for him, what with 'the cold he took, but rather I think, out of the Magnanimity of his Courage, as soon as he was put in there, he immediately dyed; by whose death I lost my Martial Mistress, but had not time to bewail it; whose death when the Prince heard of, he sent several Officers to bring me to him, and gave me his place publiquely in the Feild, with the greatest Honour and kindness imaginable: I was also that day made Adjutant General of the Army. *Major* Cheldon *dyed.*

The Foot and Cannon (by the loss of Ammunition) became wholly unserviceable to us. The Enemy persu'd, and we were to make our retreat to the *Deuizes,* otherwise called the *Vies;* Leiutenant Collonel *Nevill*[117] was Commanded to bring up the Rear, which he did with that Gallantry, and good Conduct, that we kill'd as many of the Enemy as they did of us; when the Foot came safe to the *Vies,* and that the Horse had only done that Service; instead of calling us run away Horse (which the *Cornish* used to do) they call'd us gallant Horses; for the *Cornish* Foot knew not till then the *Our Retreat to the Vise.*

Service of Horse. When we came to the *Vies,* there was found (as I heard) two Barrels of Powder, and the Bell-ropes made Matches; and 'twas Fortified as well as it was capable of, in a short time. While our Army was in and about the *Vies,* our Scouts gave no satisfactory account for several hours where the Enemy was; whereupon Sir *James Hamilton* Commanded me personally to seek out the Enemy and give a certain account where they were, or never to return again: 'Twas my good fortune in less than half an hour to fire at one of their Scouts, and by his flight to discover their whole Body, of which I gave full satisfaction.

The Enemy made no near approaches to the Town, nor did they storm it on any Part; our Horse being still in a condition to give them a Charge: Sir *James Hamilton* gave me the Honour and Trust to Quarter our Horse where I thought fit, giving him an account (at the Counsel of War, whether he was then going) what I had done: After I had quartered them in a very convenient place, I went to the Counsel of War; where I found the General Officers, and some few more at Supper; Prince *Maurice* ask'd me whether I had Quartered the Horse well, and I told him I had; he bid me say no more but sit down to Supper, after Supper, all but the General Officers and some others withdrew, and I gave my account and afterwards proffered to withdraw, but I was permitted to stay; and after the Counsel resolved what to do, I was commanded to draw up the Horse into the Market-place, where Sir *James Hamilton* received them.

11th *Sigh.* — *Lord thou hast qualified our hopes with fears! thou art gone far from us, and goest not forth with our Armies, thou makest us to turn our backs upon our Enemies, thou hidest thy Face, and forgettest our misery and trouble, thou hast compassed us about with our Enemies; But Lord arise, help us and deliver us for thy Mercies sake! for through thee will we overthrow our Enemies, and in thy name will we tread them under that rise up against us.*

THE VINDICATION OF RICHARD ATKYNS

About midnight, our Horse march'd, or rather made an escape out of Town, leaving the Foot behind us; we met not the Enemy at all, but some of our own Forces, whose fears scattered them, and we were like to fall foul upon each other: They were part of the Horse that should have come to our Assistance, but hearing ill news secur'd themselves, *viz* the Lord *Craffords*⸺[118] Regiment, between Three and Four Hundred, and Collonel *Longs*⸺[119] Regiment, between Two and Three Hundred. At the break of day, we were at least 8 Miles from the *Vies,* and free from all Enemies, between that and *Oxon;* Prince *Maurice* and several of the Officers Gallop't to *Oxon,* to be there as soon as they could; but my Horse had cast two Shoes, and I was forced to stay behind to set them at *Lambourne,*[120] where leaning against a Post, I was so sleepy that I fell down like a Log of Wood, and could not be awakned for half an hour: 'Twas impossible then to overtake them; so I went to *Farringdon,* being not able to reach *Oxford* that Night; I fell off my Horse back twice upon the *Downes,* before I came to *Farringdon,* where I reel'd upon my Horse so extreamly that the People of the Town took me to be dead Drunk: When I came to my House (for there I somtimes liv'd) I dispatch'd a Man and Horse presently to the Prince to receive Orders, and desired my Wives Aunt to provide a Bed for me; The good woman took me to be Drunk too, and provided a Bed for me presently, where I Slept at least Fourteen hours together without waking.

The retreat of the Horse to Oxford.

The next Morning I had Orders that the Rendezvous was about *Marlborough,* whether I went with Several Horse quartered at *Farringdon,* and came timely thither: The Lord *Wilmott*⸺[121] was sent with a recruit of Horse from *Oxon,* and I Suppose all the Horse at that Rendezvous were about 1800 and two small peices of Cannon; We lost no time, but March'd towards the Enemy, who stood towards the top of the Hill, the Foot in the middle between two Wings of Horse, and the Cannon before the Foot: There were four Hills like the four corners of a Dye, in such a Champaign, as 40000 men might Fight in. Upon one of the Hills we discharged our Can-

non, to give notice to our Foot that we were come to their releif:
Then Forlorn-hopes out of each Army were drawn out, and the
Lord *Wilmotts* Major, *Paul Smith* Commanded ours, who did it
with that gallantry, that he beat them into the very Body of their
left Wing, and put them out of Order; which we took advantage
of, and immediately charg'd the whole Body; the Charge was so
suddain that I had hardly time to put on my Armes, we advanced
a full Trot 3 deep, and kept in order; the Enemy kept their Station, and their right Wing of Horse being Curiaseers were I'me
sure five, if not six deep, in so close Order, that *Punchinello* himself
had he been there, could not have gotten in to them.

Engagement at Roundway-Down.[122]

All the Horse on the left hand of Prince *Maurice* his Regiment,
had none to charge; we charging the very utmost man of their right
Wing: I cannot better compare the Figure of both Armies than to
the Map of the Fight at Sea, between the *English* and the *Spanish*
Armadoes, (only there was no half Moon) for though they were
above twice our number; they being six deep, in close order, and
we but three deep, and open, (by reason of our suddain charge)
we were without them at both ends: The Cannoneers seeing our
resolution, did not fire their Cannon; No men ever charg'd better
than ours did that day, especially the *Oxford* Horse, for ours were
tyr'd and scattered, yet those that were there did their best: 'Twas
my fortune in a direct line to charge their General of Horse, which
I suppos'd to be so by his place; he discharged his Carbine first,
but at a distance not to hurt us, and afterwards one of his Pistols,
before I came up to him, and mist with both: I then immediately
struck in to him, and touch'd him before I discharged mine; and
I'm Sure I hit him, for he stagger'd, and presently wheel'd off from
his Party and ran.

My Engagement with Sir Arthur Haslerigge.

Here I must desire the Readers to be very particular in this relation, because Twenty several Persons have intituled themselves
to this Action; and a Knight that shall be name-less, that is dead
(Speaking of his great Services and small rewards to me) told me
the very ensuing story himself, all but that he could not give so

good reason as I could, why it was Sir *Arthur Haslerigge*.[123] When he wheel'd of, I persu'd him, and had not gone Twenty Yards after him, but I heard a Voice saying, *'Tis Sir* Arthur Haslerigge *follow him;* but from which Party the Voice came I knew not they being joyn'd, nor never did know till about Seven Years since, but follow him I did, and in Six score Yards I came up to him, and discharg'd the other Pistol at him, and I'm Sure I hit his head, for I touch'd it before I gave fire, and it amazed him at that present, but he was too well Armed all over for a Pistol Bullet to do him any hurt, having a Coat of Male over his Arms, and a Head-peece (I am confident) Musquet proof, his Sword had two Edges and a Ridge in the middle, and mine a Strong *Tuck;*[124] after I had slackned my pace a little, he was gone Twenty Yards from me, riding three quarters speed, and down the side of a Hill, his posture was waving his Sword on the right and left hand of his Horse, not looking back whether he were persued or not, (as I conceive) to daunt any Horse that should come up to him; about six score more I came up to him again (having a very swift Horse that Cornet *Washnage* gave me) and stuck by him a good while, and try'd him from the Head to the Saddle, and could not penetrate him, nor do him any hurt; but in this attempt he cut my Horses Nose, that you might put your finger in the Wound, and gave me such a blow on the inside of my Arm amongst the Veins that I could hardly hold my Sword; he went on as before, and I slackened my pace again, and found my Horse drop Blood, and not so bold as before; but about Eight Score more I got up to him again, thinking to have pull'd him off his Horse; but he having now found the way, struck my Horse upon the Cheek, and cut off half the Head-stall of my Bridle, but falling off from him, I run his Horse into the Body, and resolv'd to attempt nothing further than to kill his Horse; all this time we were together hand to fist.

In this nick of time came up Mr *Holmes* to my assistance, (who never fail'd me in time of danger) and went up to him with great resolution, and felt him before he discharg'd his Pistol, and though

RICHARD ATKYNS ESQUIRE

I saw him hit him, 'twas but a flea-biting to him; whilst he charg'd him, I imployed my self in killing his Horse, and run him into several places, and upon the faultring of his Horse his Head-peece opened behind, and I gave him a prick in the Neck, and I had run him through the Head, if my Horse had not stumbled at the same place; then came in Captain *Buck* a Gentleman of my Troop, and discharged his Pistol upon him also, but with the same success as before, and being a very strong Man, and charging with a mighty Hanger, storm'd him and amaz'd him, but fell off again; by this time his Horse began to be faint with bleeding and fell off from his rate, at which said Sir *Arthur, What good will it do you to kill a poor Man,* Said I *take quarter then,* with that he stopt his Horse, and I came up to him, and bid him deliver his Sword, which he was loth to do; and being tyed twice about his wrist, he was fumbling a great while before he would part with it; but before he delivered it, there was a run-away Troop of theirs that had espied him in hold; Sayes one of them *My Lord General is taken Prisoner;* Sayes another, *Sir Arthur Haslerigge is taken Prisoner, face about and Charge*; with that they Rallied and charg'd us, and rescued him; wherein I received a Shot with a Pistol, which only took off the Skin upon the blade bone of my Shoulder,

Sir Arthur taken Prisoner.

Sir Arthur rescued.

Hos ego versiculos feci tulit alter honores.[125]

This Story being related to the late King at a second, or third hand, his Answer was, *Had he been Victualled as well as fortified, he might have endurd a Seidge of Seven Years,* &c. His Horse died in the Place, and they hors'd him upon another, and went away together. 'Twas one of the best Horses the late King had at the Mewes he rid upon, and 'twas the late Kings Saddle, which I had; and when I went to the Parliament Quarters, I gave it to Sir *Henry Wroth:*[126] When we came back to the Army (which in so confused a Feild was difficult to do) we found the Enemies Foot still in a close Body, their Musquets lin'd with Pikes, and fronting every way, expecting their Horse

The defeat at Roundway-Hill.

to rally and to come to their relief; in the mean time our Horse charged them, but to no purpose, they could not get into them; at last, when they saw our Foot March from the *Vies,* and come within a Mile of them, they ask'd quarter, and threw down their Arms in a moment: We lost few men (especially of quality) and they many; Sir *James Hamilton* was very much hurt, and Leiutenant Collonel *Molesworth* my Leiutenant Collonel, who went to *Oxford;* by whose absence I commanded the Regiment in cheif: When I alighted I found my Horse had done Bleeding, his cuts being upon the Grisly part of his Nose, and the Cheek near the Bone; I bid my Groom have a great care of him, and go into quarters immediately, but instead of going into Quarters, this Rogue went directly to *Oxford,* left my hurt Horse at *Marlborough* with a *Farrier,* and sold another *Barb* of mine at *Oxon,* and carried my Portmanteau with him into the *North,* which had all my Clothes and Linnen in it, and other things worth above 100*l.* and I never saw him more: My charging Horse I had again, and some Mony for my *Barb,*[127] which was bought at an under value, being no good Title: But for want of a Shift (my Wound having Bloodied my Linnen) I became so lowly in three or four daies, that I could not tell what to do with my self; and when I had got a Shift, which was not till we took *Bath;* my Blood, and the Sweat of my Body had so worn it, that it fell off into Lint.

—O Lord thy Wrath endureth but the twinckling of an Eye! Sorrow may endure for a Night, but Joy cometh in the Morning: Thou hast not shut me into the hands of the Enemy, but hast set my Feet in a large Room; thou hast been my hope and a strong Tower for me against the Enemy: The Lord is on my side, I will not fear what man can do unto me: Through God shall we do great Acts, and it is he that shall tread down our Enemies; the Lord of Hosts is with us, the God of Jacob is our refuge.   *12th Sigh.*

Soon after the Prisoners and Trophies and the Matters of the Field were disposed off, a Council of War was called at the *Vies,* and Prince *Maurice* sent for me thither to attend him, as soon as

RICHARD ATKYNS ESQUIRE

Prince Mau-
rice *his kind*
intentions
towards me.

I was come he had notice and came out to me, and bid me wait there, telling me, that he intended to send the Trophies, and Letters to the King by me; but upon comparison of Commissions, 'twas found that the Lord *Wilmott* was Commander in chief for that expedition, who sent Sir *Robert Welsh,* and I returned to my Quarters. After we were refresh'd in our Quarters, we March'd to *Bath;* which Town the Enemy had newly quitted, and March'd, or rather retreated to *Bristoll;* there I found my Leiutenant *Tho. Sandys,* formerly taken Prisoner, recovering of his Wounds, but not well able to go abroad: We were very glad to see each other, and I desired him to tell me the manner of his being taken; he told me, he persued the Enemy too near their Body, and they sent out fresh Horse upon him and took him, and gave him Quarter without asking; but after he was their Prisoner, a *Scot* Shot him into the body with two Pistol Bullets, which were still in him, so that he was very near death: That when he was brought to the Town, Sir *William Waller* enquiring what Prisoners were taken, heard of his name, and came to see him; he seem'd exceeding angry at the inhumane Action that befel him, and sent for his own Chirurgion immediately, and saw him drest before he went away; he gave the Innkeeper charge that he should have what ever he call'd for, and he would see him paid; that whatsoever Woman he sent for to attend him, should be admitted, and lent him Ten broad Peices for his private expenses; and before he March'd to *Bristoll,* he came to see him again, and finding him not able to March, took his Parole, to render himself a true Prisoner to him at *Bristoll* when he was able to ride; which I found (for his word sake) he was inclinable to do; to which I answered, that he was now made free by as good Authority as took him Prisoner, and that I expected he should return to his Command; upon which we struck a heat and at last referred the business to the Lord *Carnarvan* to determine.

Sir William
Waller *Hu-*
manity.

The Case we agreed to be *Whether a Prisoner upon his Parole to render himself to the Enemy, being afterwards redeemed by his own Party, ought to keep his Parole or not.* His Lordship heard Arguments

THE VINDICATION OF RICHARD ATKYNS

on both sides; at laſt said thus, that there had been lately a presi- *A Case in*
dent in the Council of War in a Case of like nature, wherein it *War deter-*
was resolv'd, that if the Prisoner (being redeem'd by a Martial *mined.*
Power without any consent of his own) shall afterwards refuse
the command he was in before, and attempt to render himself
Prisoner to the Enemy, he shall be taken as an Enemy, and be
kept Prisoner by his own Party; the reason seems very ſtrong, because he may be prevailed upon by the Enemy to betray his own
Party; and the freeing of his Person, gives him as it were a new
Election; and if he choose rather to be a Prisoner than a Free-
man; it demonſtrates his affection to be there. But this did not
satisfie my Leiutenant, for he would not take his Place as before,
but March'd along with the Troop as my Prisoner, till the taking
of *Briſtoll* (the place where he promised to render himself) and
then he thought he was fully absolv'd from his Parole, and betook himself to his imployment again.

When we came to *Briſtoll*, Prince *Rupert* (whose very name was
half a Conqueſt) with the *Oxford* Army lay before it on the Weſt
side, and Prince *Maurice* with the *Weſtern* Army on the Eaſt: Both
Armies being not half enough to beseidge it; our *Cornish* Foot were
to fall on firſt, which they perform'd with a great deal of gallantry
and resolution; but it proving the Stronger part of the Town, they
were beaten off, with a great deal of loss; When they found it in-
accessible, they got Carts laden with Faggots, to fill up the graft;
but it being so deep, and full of Water (though they attempted it
again very gallantly) they could do no good upon it; but as gallant men as ever drew Sword (pardon the Comparison) lay upon
the ground like rotten Sheep: Amongſt which were slain Collonel *Brutus Buck*, Sir *William Slaning*,[128] Collonel *Trevanion*,[129] with
many more, howsoever this loss of ours drew the Town Forces that
way, which might be some advantage to Prince *Ruperts* Forces,
who ſtorm'd that part of the Town with such irresiſtable Cour- *The taking of*
age, that forced them from their Works, and gave admission to *Briſtoll.*
the Horse; who soon beat them from their Guards, and took the

RICHARD ATKYNS ESQUIRE

Town; by taking of which, the West was uqpon the matter cleared, for this gave such a reputation to our Army, that made all the Garrisons thereabout willing to come in upon Terms, as *Dorchester* and *Poole* came in to the Lord *Carnarvan,* who was sent with a Party of less than a 1000 men, to demand them; both which rendred within one Week. I cannot tell whether the Articles were too large; but I too much feared the not strictly performing them, did moulder two gallant Armies; the one before *Lime,* the other before *Gloucester,* that otherwise might have come in; which had they taken, or the Field Army March'd up to *London,* to have joyn'd with the Marquess of *Newcastle,*[130] who was then successful in the *North,* there had been probably, an end of the War on the Kings side; but this is not to the purpose in hand, for which I ask Pardon, and return to my self.

The taking of Dorchester *and* Poole.

When we were possest of *Bristoll,* and the lesser Garrisons came tumbling in to the obedience of the King, I took the Kings Crown to be settled upon his head again; and my place of Major to the Prince, being supplyed by a more knowing Officer, I desired leave to return to my private condition as before, and to march with the Army that was to beseidge *Gloucester,* hoping to possess my Estate there; offering my Leiutenant, Cornet, and several other Gentlemen the Command of my Troop, all which refus'd to serve in my place; so that it ceast to be a Troop, and became a little Army, for in less than a Week, I preferred Twenty of them to be Captains, Leiutenants, and Cornets.

The Seidge of Gloucester.

After the Seidge of *Gloucester was rais'd,*[131] the Parliament that were very low before, began to prick up their ears again; and I retir'd to *Oxford,* where I was offer'd several very good Commands, but refused them all: At length Mony began to grow short, and my Wives Estate and mine, were chiefly in the Parliament Quarters; my Wife therefore went up to *London* to make Mony, there being several Hackney Coaches come down with the Commissioners, which were to treat at *Oxford,* which had the Parliaments Pass, to go up and down; by which means, I thought she might go safely;

THE VINDICATION OF RICHARD ATKYNS

yet I endeavoured to get a Pass of the Lord General also, but could not obtain it so soon as the Coach was to return; so she adventured without a Pass, which proved very unhappy, for at *Nettlebed,* a party of Sir *Jacob Ashley's*[132] Souldiers (who was then Governour of Reading) took her Prisoner, and carried her to *Reading,* where the Waters being then overhigh, she took a great fright. When she was brought to Reading, Young Sir *John Cademan,*[133] ask'd the Souldiers who she was, and when he understood she was my Wife, he us'd her with all the respect that could be, and waited on her to the Governour, who for my sake exprest many civilities to her; and sent a guard with her into the Parliament Quarters.

My wives unhappy Journey to London.

When she came to *London,* she found her House full of Souldiers, and could not be admitted there, but lodg'd at the House of one of her Tenants that joyn'd to it; where seeing her Goods carried away before her Face, and finding her Uncle (who was possest of her Estate) inhumanely unkind to her, she fell ill and miscarried (as I understood) of three Children, and was very near her death; yet in this Sickness she sent a Token to me, by a kinsman to *Oxford,* and desired a Lock of my Hair, which she had. Here I must not omit to tell you, that though her own Uncle us'd her most unlike a Christian, yet my Uncle us'd her as his own Child; from whom she had advice and protection, or any thing else she wanted. After, I thought fit to remove to *Bristoll,* and she unsatisfied with my absence, sent a Gentlewoman of quality to see how I did there; who was forced to go the greatest part of the way on Foot, the War being then so hot, that men could not Travel. After half a Years stay at *Bristoll.* I came to *Oxford* again, towards the the Treaty at *Uxbridge;*[134] whether the King sent me with Letters to his Commissioners: My Wife retiring thereabouts for Ayre, heard of it, and came to me, but yet so weak, that when she alighted out of the Coach, she fell into a Swound, and came not to her self in half a quarter of an Hour: There I staid as long as the Treaty lasted; being the very last Messenger that was sent back to *Oxford;* and I remember the Message was, that the Parliament Ministers were

Her great kindness to me.

now upon proving *Presbitery* to be *Jure Divino*.[135] After that Treaty, I staid not long at *Oxford*, having no imployment in the Army; but ask'd the Kings leave to go up to *London*, who did not only give me his free leave, but withall, directions how I should carry my self there, to this effect, *viz.*

The Kings gracious advice.

I know you are too much a Gentleman, and my Friend, to do me any prejudice; and if you will do me any good, it must be by getting an interest in your Friends and Tenants, and putting Mony in your Purse—: I know there will be those that will arise in my behalf, but they will be so small and insignificant, that it will but establish my Enemies the better; by no means joyn with them, but the time will come, in which my People will return to their obedience; and though perhaps I may never see it my self, I am confident my Son shall.

And so gave me his hand to kiss, and the Lord of *Bristoll* then Secretary, gave me a Pass, and Collonel *Legg*, then Governour of *Oxford*, gave me direction which way to go, having also a Pass from the Speaker of Parliament.

But before I leave *Oxford*, I must acquaint you with one passage there; though I have ever observed *Tullyes* rule, that true Valour is neither to do, nor suffer injury, and have avoided Duels as much as in Honour I could (rather to preserve a good Conscience, than an ill Carcase) yet some I have been very near, of which I will onely name one.

A Challenge.

In *Oxford*, a Knight provoked me with so ill Language, that I could not forbear striking of him; and being very angry, I took his Perriwig off from his Head and trampled it under my Feet: The next morning he sent me a Challenge by his second, a person of Quality, who found me in Bed; I desired him to stay a little, and I would send for my second, to go along with him to his Friend, which he did; when I sent my Servant for my second, I also commanded him to secure a good Charging Horse by the way, (intending to Fight him on Horse-back;) in less than half an Hour, my Second and Servant came to me, and then (having the priviledge to choose

THE VINDICATION OF RICHARD ATKYNS

Time, Place, and Weapon) I told him I would meet his Friend and himself, with this Gentleman my Second, on Horse-back in *Bullington Green*, between two and three of-the Clock that afternoon, with Sword and Pistolls, and without Arms, and shewed him my Sword and Pistols; saying withal, I would not except against any his Friend should bring; which he desired me to send in Writing, and I did it by mine own Second, who brought me word back, that all was well accepted: We all din'd together with other Company (as we us'd to do) without any suspicion of a Quarrel, that I know of, after dinner, my Second and I mounted, and as we past by *Magdalen* Colledge (which was the way to the place appointed) the Second on the other side, call'd to my Second, desiring to speak with him, his business was to perswade me to alight, and treat of the matter in his Chamber there; they both intreated me to alight, assuring me 'twas nor dishonourable, but fit so to do; but I not convinced by their arguments, utterly refus'd, unless the Principal himself would come and desire it, and absolve me from my promise of meeting him at the place appointed; which he did, and then we went up together to his Seconds Chamber, where we found an Earl (whose Sister the Knight had Married) who also pretended Friendship to me; he urg'd how much I was before hand with his Brother, and propos'd as an expedient, that I would declare I was sorry for what I had done, which I desir'd to be excused in; at last he offered that if I would say I was sorry for what I had done, he should say he was sorry for giving the occasion; which was acknowledged on both sides, and so we were made Friends. *A Duel prevented.*

13th Sigh.

—*Lord how various are thy dispensations, thy wayes are past finding out; thou hast turned our Sorrow into Joy, that part of the Wheel which was lowermost, is now got uppermost; thou hast turn'd my Sword into a Plough Share, and my publique dangers and wants, into private Peace and Plenty; and instead of the cold Mother Earth, and Canopy of Heaven, thou hast prepared a warm and pleasant Bed, Lord grant me Grace suitable to thy Mercies.*

RICHARD ATKYNS ESQUIRE

From *Oxford,* the first night I came to *Uxbridge,* where I thought fit to acquaint the Governour of the Town with my Condition, least I should be taken as a Spie; but not-withstanding my Pass (to shew his Officiousness to the Parliament) he carried me Prisoner to *London,* where I was forced to make Friends to the Speaker to be released: Soon after, I attended my Composition[136] at *Goldsmiths-Hall,*[137] where (though I was a Prisoner to the Parliament) yet my Creditor arrested me upon an Execution, and carried me carried me away to the *Poultry Compter,*[138] and then sequestred my whole Estate; which if he had done before, might have paid him half his Debt. There I staid about, a Year without seeing a Yard of growing Grass; striving to be releast upon the Priviledge of Parliament, being their Prisoner; and therefore could not in Honour be a private persons Prisoner; but that failing me, I removed to the *Kings Bench,*[139] where (the Lady *Lenthall*[140] being my Kinswoman) I had convenient liberty: Whilst I was in Prison, my Wife agitated my Composition, rais'd mony and paid it, and I wanted not; moreover I put her upon the taking off 300*l. per ann.* settled upon her, after Marriage, in exchange of Lands between us (though she had no mind to attempt it) and she clear'd that from Composition; which several Ladies urg'd, as a president, but was not granted to any: She was then contented to release these Lands for the Payment of Debts; She having her own again, which if they had been Sold at a full Value, would have paid all.

In the mean time, I was not wanting in retaliating kindnesses; for by my diligent soliciting for her, and laying out a good part of the Mony I received in sale of my own Estate: I got the greatest part of her Estate out of her Uncles hands, into the hands of Friends fitter to be trusted; and paid Debts and Legacies charg'd upon her Estate with the rest; and then conveyed her whole Estate in *Middlesex* and *Berkshire* (whereof a considerable part was Lease Lands and so mine) To Trustees of her own choosing, acknowledging all valuable considerations, and using all barring words that council learned could invent, to avoid fraud; not in satisfaction of what

The cruelty of those times.

A good wife.

A good husband.

I owed, but rather in the confidence I had in her, that what she enjoyed, I should never want; but besides my confidence of her love, that which induc'd me to it was, that no man of my condition could call any thing his own; for there followed Decimation, Imprisonment, and what not; and who was more fit to be trusted than such a Wife, who seem'd to be altogether compos'd of Kindness, and of whom I had not then any cause of Suspicion; yet I observ'd that she was very strict and over curious in the Penning of her Conveyances, thinking that she had never enough, and ever suspitious of her best Friends, adhering rather to those that had a design upon her, and did admire her Wit and Parts, than those that told her the plain truth, and wish'd her happiness really.

Whilst I was in the *Kings Bench,* my Chief Creditor, who Arrested me in Execution, lead me out of the Rules, to treat of the Debt between us; which the Deputy Marshal perceiving, said, he was glad to see us agreed; my Creditor answered, we were not yet agreed; said the Marshal, you must agree now, for this is an escape in Law, and he may chose whether he will go back to Prison again; which amazed my Creditor: But I told him, 'twas not for my Honour to revenge my self upon him, and use him as he had us'd me; I would only take the advantage by this, to make him honest, and to abate the extraordinary Charges he demanded; but his Principle and Interest he should have to a farthing: Which I paid him within one Week, and all my own Debts; there were also several Debts I was engaged for with my Uncle *Sandys,* entred against me; which the Creditors (out of the confidence they had of my justice) withdrew, and took my Word for their Payment; and I Paid them all, within one Years time. *A great Temptation.*

Being now in some measure free from Debts, my Wife and I retired into the Country; where we found the common Fate of a sequestred Estate; not onely a ruin'd House, but Gardens and Orchards ploughed to the very Doors: Yet with all its faults, we liked it better than any place we had in the Country; being also encouraged by the kindness of my Neigbourhood to live there (though it were a Lease holden of the Dean and Chapter of *Gloucester*). Too soon after Church Lands were exposed to Sale by the then usurped power, calling themselves a Parliament; and upon consideration of all things, *viz.,* that if we did not Buy it a Souldier might; who would immediately Cut down the Timber, with which it was well stor'd, and force me out of the term had in it already (which was Eleven or Twelve Years to come) besides, I thought it a point of Conscience, to preserve the Church Lands to my Power; and it was the principal seat of my Name and Family for above a Hundred Years: Upon these thoughts, my Wife and I Sold Lands, and joyned to buy the reversion after the Lease in Trustees names; and I was so kind to her, as to make it over in Joynture, in case she outlived me, which in Common Barter, was at least three times worth the Mony.

Mutual kindness.

During the Kings absence, I cannot say I liv'd, but breath'd only; being altogether a Patient, and not an Agent: Yet even then there were some remarkable passages, of which I will name but one.

A passage at a Committee

Upon review of Sequestrations, I was Summon'd to *Hickes's-Hall*,[141] to compound for my Wives Estate in *Middlesex,* in answer to which, pleaded a Lease of 21 Years, made by my Wives Father, Anno, 1620. for Payment of Debts and Legacies, which were not then paid, and the Lease still in force, to which, when they could make no reasonable objection; said *Barksted* (then Leiutenant of the Tower)[142] Sir, *do not you live with your Wife*, to which I answered, *Yes Sir that I do, and I hope I may without offence;* Said he, *that's cause enough for Sequestration,* to which I reply'd, *I hope this Honourable Committee doth not sit here to part man and Wife;* sayes another of the Committee, *What Sir do you accuse us for parting Man and Wife;* to whom I replyed, *I say I hope you do not;* with that, I was bid to withdraw, and they past no Judgment upon it that day; and afterwards I applied my self to the Chair-man, who freed it from Sequestration; I was so plac'd by fortune, that I could not move without discovery; so that I had not power to do more, than to advise my Friends and Relations to obey the King and His Laws, and to make

Mr. *Henry Norwood*[143] Executor of all my Arms, who us'd them better than I could have done my self.

Not long after, it pleas'd God to restore the King and the Church again, before the Lease formerly mentioned was quite expired; by which means, my Wife and I lost our purchase Mony, and the repairing of the House, improving of the Grounds cost at least 1500*l*. to make it a Seat; Which when I apply'd my self to the Dean and Chapter to renew; I found the Chapter generally my Friends, and two of them being authorized to treat with me concerning it, offered to leave the Fine to my self, so I would increase the rent 10 *l. per ann.* more, which offer (though it were very obliging) yet I desired to consider of it, (the increase of Rent being new to me) whereupon the Dean took the matter into his own hands, and then I found a great alteration; for whereas the Gentlemen that treated with me before, would in kindness come to me, or meet me; I could hardly speak with the Dean, having (as I was inform'd) a design to have it himself, which troubled me very much; but my old Friend Doctor *Labourne* coming to visit me, found me discompos'd, and ask'd me the reason of it; which being as I conceived, to the Scandal of our Church, I told him 'twas out of his Road to help me; which made him the more importunate, saying, I knew not his Power; At last I told him that I was oppos'd in the renewing of a Lease I had holden of the Dean and Chapter of *Gloucester;* he said, he had seen the Dean beyond Sea at Queen-Mothers Court, and made no question, but to make him my Friend; and went immediately to the Countess of *Guilford,*[144] and she had a Command from the said Queen to sweeten him; who did it so effectually, that whereas, he would not renew it upon any terms before, he then agreed upon a reasonable Fine; and was so wholy for me, that if I did not take it in mine own name, and to mine own use, (Exclusive to my Wife, and all others) there should be a bigger Fine set upon it, I did accordingly pay the Fine, was continued Tenant, had my Lease, and he staid his stomach upon another Mannor.

The Prebends worthy Persons.

A true Friend.

RICHARD ATKYNS ESQUIRE

After the Kings blessed restauration, I began to be an Agent again, and fell into several great and chargeable Suites *viz.* against the Company of *Stationers,* to Vindicate the Kings Legal Right in Printing the Common Law against Sir *Robert Vyner*[45] and several others; of which should I give an account, this would be rather a Volume than a Treatise. The taking out of the Depositions of one Suit only, amounting to almost 20 *l.* Yet this I can most truly say, I never began a Suit, but upon necessity; and asking at *Abel*[46] first: And though I prevail'd in them, yet they cost me so much Time and Mony, that I may well say as that General did; who after a great Victory, upon the veiw of his Army, found it was obtained by the loss of so many Gallant men, that he said with Sorrow to his Officers that remained; *Many such successes will undo us.* So these fresh expences, together with paying a Fine for what I had bought before; being put into Honourable, though Chargeable imployments, as Deputy Leiutenant, Justice of the Peace, an Officer of the Militia, and living in a more plentiful manner than before, to entertain my Friends, and having no help at all; I became in Debt again, and borrowed 1500 *l.* upon the said Colledge Lease; with which I paid some of my Wives Debts, as well as mine own, alwaies believing I should never see the day wherein our Persons, or Estates should be two.

14th Sigh.   *Lord thou hast made mine Enemies rejoyce over me, and laugh me to scorn! thou hast set me fast in Prison, yet thou hast not given me over, but hast sweeetened it with the Kindness of my neerest Relations, that with* Paul *and* Silas, *I have enjoyed many pleasant hours there; thou hast put mine Adversarie into mine Hands, and Blessed be thy Name that thou hast restrained me from requiting Evil for Evil; and though I have not arriv'd to that height of Piety as to Love my Enemies, yet 'tis a great comfort to me, that I have made so good a beginning as to be just to them, as well as kind to my Friends.*

Here I must crave pardon of the Reader to fall from a general discourse of my Life, to a particular vindication of my self, wherein

THE VINDICATION OF RICHARD ATKYNS

I am falsely accus'd; which indeed is the cause of this Treatise, and though I tremble to do it, yet I choose it as the least of two Evils, being not able to undeceive the World by any other means.

After my Wife had found the sweetness of enjoying a quieted Estate, she began by degrees to decline my advice, and keep all things private from me, and giving ear to Adulators; she was miserably cousened, and bought all things at the worst hand, and run in Debt to most Tradesmen that would trust her; which when I perceived, and was dun'd for the Mony, I prevail'd with her to shew me her accounts; upon perusall of which, I found many Errors, *viz.* no perfect Rental whereby to charge the Receiver, the Accompts false cast up, and several considerable Sums doubly charg'd; But when I told her of it, and desired these Accompts might be evened, before she trusted him any further; the result was, that I should never pry into her Accompts more, for there being some private Reckonings between them, he had the power to make her set her hand to any thing, which she hath dearly paid for since; and though she exclaimed against him, that he had let Leases to the Tenants of more Ground, and for a longer Term than was agreed, yet she ever Seal'd the Leases, and most commonly without Counter-parts. By this ill management of Affairs, several Debts were contracted and my Person liable to them all; this I was very sensible of, but how to remedy it I knew not; for the staff being now in her hands, and out of mine, she might as well beat me as assist me with it. *The ill consequence of Debts.*

The danger of Trusting.

Yet being provok'd by a Conjugal duty, I took the boldness to tell her, we grew more and more in Debt every Year, which would be destructive at last; that the best way would be to Sell and pay them off, and live as Comfortably as we could upon the remainder, or to let the House, (the keeping of which, much increast our Debts) and live privately for a time: Neither of which she liked, but the latter she could never forgive, to grudge her a House to Live in (as she often repeated) and this was the first real difference I can remember between us; as if Persons indebted by keeping one House, *Plain dealing out of fashion.*

RICHARD ATKYNS ESQUIRE

could pay their Debts by keeping two: this was bandyed about every where, and gave an occasion to rip up all my other Faults, complaining that I never lov'd her; and as an argument urg'd, that I left her company, and went into the Kings Army; which to the Zealous Ladies, who like Tinder, take Fire upon the least occasion; and would have none go that way, was fault enough, without any other to be condemned, forgetting as (Zealous as they were) *Uriahs Answer to King David,* when he would have had him gone to his House, *&c. The Ark, and* Israel *and* Judah *abide in Tents, and my Lord Joab, and the Servants of my Lord are incamped in the open Fields; Shall I then go into mine House, to Eat and Drink, and lye with my Wife,* &c. *I will not do this thing,* said he: so may I answer, shall the King, the Nobility and Gentry, and all be concern'd for their Lives, and I be in a Voluptuous Bed my self! certainly 'twas not to be done by a good Subject: And this I urge the more home, in regard her fondness was for above 20 Years together, as great as her hatred can be now, for I could not be out of her sight (though about her own business) without strick't Examination; sometimes smelling my breath, whether I had drank Wine, and my hair, whether I had taken Tobacco; and though I never could abide Tobacco in my life, yet if she had smelt it in my hair, by being where it was taken, the World could not perswade her to the contrary, but that I had taken it my self.

And here I cannot forbear to tell you a passage. About Twelve Years since, a Noble Person and his Lady Din'd with us; after Dinner, he whispered me in the Ear, which my Wife perceiving, acquainted his Lady, that there was something extraordinary to be done; so both the Ladies came to kiss the secret out of us, and when 'twas discovered, 'twas no more than this that there was a Musick meeting at Master *Lawes* his Chamber that Afternoon, and if they pleas'd to go we should be glad of their Company; which they (otherwise engag'd) refus'd to do: As we went, sayes this Person to me, let us endure these impertinencies in our Wives; for when women begin to be indifferent, they begin to hate: Which saying

2. Sam. Chap. 21.

Extreams never last.

hath prov'd most prophetical to me; for she having the sole power of an Estate, and being the only Child of her Father, and so fonded by him, that none must displease her; whereby she had her will in all things, and being in her nature passionate and high, which by her Education, was much increas'd; this Settlement could not please her, but she must needs have all that remained also, to render me worse than her Ward, for though a Ward hath nothing, yet he owes nothing, and is maintained by his Guardian; but I was clog'd with mine own and her Debts which lay hard upon me, yet must not have wherewith to pay them, but like the *Israelites,* make Brick without Straw. *Fondnesss an Error.*

Here I must take leave to observe, that the great reason why the Law favours Marryed Women so much is, because 'tis presum'd the Husband is possest of the whole Estate, both real and personal; and therefore whatsoever Debts the Wife contracts, though never so ill, chargeth the Husbands person and Estate; which makes my case worse than ordinary: For though my Wife be (by my indulgence) as a Husband, in respect of the Estate, yet is she as a Wife, in respect of the Debts she owes; having the benefit of the Estate, and not answerable for the Debts by Law; which not onely contradicts reason, but the very Creation it self: *For Man was created in Gods Image, to be ruler of the whole Earth, and woman was afterwards created to be a help meet for man:* So man was made for Gods sake, and woman for mans sake. But did she perform accordingly? Was she a meet help? I doubt upon examination, you will not find her so; for, *the very first Act she did, was to give her Husband the forbidden Fruit, whereby he became naked*. And this was my fault, for I did all things my Wife desired me, till I divested my self of that power the Law gave me, and became naked: Though to effect this, she us'd Arguments as plausible as *Absolom* did when he kist the men at the *Kings Gate*, *what Justice he would do when the power was in him*. But unnatural power is most commonly destructive to the receiver as well as the giver; 'tis only happy when both make but one power, since those that contend for it most deserve it least: Yet to please her, I forsook   Gen. *Chap.* 1. 2. 3.   2 Sam. *Chap.* 15.

my reason and yeilded, that if she would assure me both our Debts should be paid, I would make over all to her; which when she did not approve of upon those terms, nor give me any satisfaction why she did not, but expected I should have an implicite faith in her, which she allows not to the Church; I concluded she resolv'd to blast my good name, and make me a perpetual Prisoner, which indeed was the designe; for fond love being now turn'd into deadly hatred, she spent her whole time in getting a party privately, and alienating my Friends and Allyes from me upon untrue pretences, perswading my neerest Kindred and Relations, to joyn with her and desert me; but they having more discretion and integrity, then to joyn in so unjust a Cause, rather advised her to a Composure; and though she took their Councel in other things, yet she forsook them in this, and went to another Counceller of Law, wholly unconcern'd, who also advis'd her to a Complyance; had she read the words of Matrimony, and weighed her promises thereupon, she could not need farther advice.

But when she found honest Lawyers, not for her turn, she called in Rumpers[147] to advise with, and her flatterers (who had been Pentioners to her for several Years) of *England, Scotland* and *Ireland;* the Chief of which, was her first Husbands supposed Neece, who heretofore was forbid by several Letters from her self, to come out of *Scotland* to her, but her love (otherwise call'd her wants) necessitated her to thrust her self upon her; who being a Lady of Fortune, and well furnish'd with *Scottish* Magick, she was Captain of this black Guard, who designd to build their Houses out of our ruins; but these were only preparatory for her intended separation, being only Women of as little Credit, as Estate; whose business it was to pry into my Actions, and to tell my Wife what they observ'd to my prejudice, which they found very pleasing to her: *Reproaches.* If I did Sweat going about my business, all the House must ring of it; as if it were a Miracle for a fat man to Sweat; if she were inform'd I was in company of any woman (though untrue) 'twas so great a fault as not to be forgiven. These, with matters of a higher na-

THE VINDICATION OF RICHARD ATKYNS

ture, were underhand infus'd into the Neighbours, and Tenants, to render me odious to them; as that I had spent her Estate, when indeed I got and preserv'd it with the loss of my own; and she still lives plentifully in the mid'st of it, without any contradiction, and is possest of all, but what was sold for Payment of her Fathers Debts and Legacies: That I had some strange and secret wayes of spending my Mony; whereas in truth, I spent most of what I had in *Westminster-Hall;* cheifly for the defence of her Estate, which was alwaies litigious, or in the House we liv'd in. But fears and jealousies (which were the causes of the late Wars) were also the causes of my misfortunes for this *Scottish* Lady, finding my Wives love in the Wane, and her inclination to take all advantages against me; made it her business to engage all the Servants in the House of her side, Rooting all out that would not take part with her, except my own menial Servant; who had a weary life, because he would not betray me; Notwithstanding which my Wife by fits would return to her Kindness, and (to give her her due) did many times assist me with Mony at a pinch; at which the *Scottish* Lady (*Judas* like, keeping the Purse) was much offended, saying sometimes, that if my Wife were rul'd by me, she would be utterly undone; and at other times, when I was in the Country, that she hoped never to see me there more, for my Wife was never well when I was with her: These things, when my Wife was acquainted with, she seem'd not to believe, nor redress; and then, and not till then I found I was supplanted.

More reproaches.

Judas had an excuse.

—*Lord stand up in thy wrath, and lift up thy self because of the Indignation of mine Enemies! turn thee unto me and have mercy upon me, for I am desolate and in misery; O bring thou me out of my troubles! consider mine Enemies how many they are, and they bare a Tyranous hate against me; but let them be ashamed and confounded together that seek after my Soul to destroy it; let them be driven backward and put to rebuke that wish me Evil; For the Lord is my refuge, and my God is the strength of my*

15th Sigh.

[107]

Confidence; he shall recompence them their wickedness, and destroy them in their own malice; yea, the Lord our God shall destroy them.

By this time her using of means to be separated from me began to be no secret, for she did not onely use the Instruments aforesaid in buzzing into the Neighbours and Tenants ears, how unkind and unjust I was to her privately, but profest it openly at Table, to make me cheap with my Friends and Servants; insomuch as I have several times with silence risen from the Table before I had half Din'd, and insinuated it, into Persons of the best Quality; and that it might the better take with them, she declared how ill a Husband I had been of my own Estate, and hers; how vastly I was still in Debt, and how necessary 'twas now to look to her self, to preserve a lively hood for both (pretending still some kindness to me) and that she did it for my sake, to make a party out of my own Kindred against me) and that she had already sold much of her Estate to pay my Debts; Whereas all the Estate she ever sold in relation to me (for the rest, was in satisfaction of Debts and Legacies given by her Father) was some Land at *Farringdon,* for 1250 *l.* of which about half was by her own desire made use of, with some more of mine, to buy the Reversion of the Lease aforesaid, and we both lost our Mony, which was a general case.

But she not contented with any loss wherein I was concern'd, would have had the Lease renewed in other mens names, which to please her, I propos'd to the Dean; his answer was, that if I took it in other mens names, he would put another price upon it, in regard the Chapter took so small a Fine of me for having suffered so much for the King and Church. So I took it in mine own name, but she, seldom satisfied with any reason, but where her will is predominate, never left off, till she oblig'd the Dean to promise, that upon the next renewing, it should be to Friends in Trust for her; and made use of that unlucky person (who hath been so fatal to me, and all his own Name and Family, by being engaged for his Debts) as a proper Instrument to effect her design.

What Evils attend hatred.

One Evil begets another.

THE VINDICATION OF RICHARD ATKYNS

When I desired to treat with the Dean upon a second renewing, he refus'd to treat with me, but referred me to this man who agreed with the Dean for 250 *l*. on my behalfe, which was afterwards made a Chapter Agreement, as he confessed by his answer in the Exchequer; (whether I was forced to fly for repair;) for though I gave notice that my Mony was ready, yet the Dean shuffled it off, and declar'd that unless 'twas renew'd to Trustees for my Wife, it should never be renew'd: So that this is the *English* of it, I must pay the Fine for the Estate, and she, or some others, must have the benefit of it, or it must not be renew'd at all. But this was not possible to be done, in regard my necessities compelled me to take up Monies upon it, without the payment of which, I had not power to do it; yet this I offer'd, that if she would pay my Debts, I would trust her with it; but nothing would satisfie her, but defrauding my Creditors, and leaving my Person in danger; this being all I could properly call my own to pay my Debts with, and give me a subsistance. *Some women never satisfied.*

My refusing to do this in her own way, was so great a fault and so ill taken, that it put her in King *Ahab*'s Distemper, and I had no manner of quiet for Six Months together, at Bed, or Board: Sometimes complayning to my nearest Kindred of me, and if they would not take part with her, she would discountenance their Visits; othertimes keeping me waking whole nights, thinking by her importunity, to make me the unjust Judge; when at the same time she pretended Sickness, as she did more or less for the whole Winter together (though she were as well as in any Winter I can remember) and neglected her own business wholly, to compass this design, and nothing I did or said could please her.

She was also very much pleas'd at my misfortunes, and when I complain'd to her of two sad hearings I had in Chancery that Morning, I never saw her merrier, than at that Dinner; and when I was Arrested for her Debt, (which indeed I knew not of) she seem'd more troubled that I was got off, than that I was Arrested, though at the same time she was very passionately concern'd for *Revenge is sweet.*

a Chare-woman she kept in the House, who, for forswearing her self, was Arrested and carried away, when I had the Gout in extremity upon me, and was very angry with me that I would not be so too; the rest I am ashamed to declare. When I importuned her to entertain a Kinsman of mine, whose civility might justly expect a treat for himself and his Friends I onely desir'd the use of my own Plate for that day, assuring her she should have it again, and that she should be at no charge; I could not for my life obtain it of her; nor was this all, for in further pursuance of her design, she broke a Door into the Neighbouring House; and whereas a Kinsman of mine and my Servant lay in that Chamber before, they were removed from thence, and her sweet Neece (as she calls her) lodg'd in their Room, to hold correspondence with them, and withdraw their kindness from me; and as I suppose, to remove Goods thither, as there should be occasion. And till this time, I thought the Inhabitants of this House the most impartial well meaning people to us both that could be, and there was no person that profest more friendship to any man than the Master of the House did to me; which caus'd me to make him the very Cabinet of my Secrets; and because I had once those apprehensions, I shall leave out much of what I can say for my advantage, till there be a more clear discovery of the truth as touching him, or a publick denial of the truths herein asserted; for which I shall reserve a reply, if any such shall happen. However this Door, which I hoped had been made for both our accommodations, to give us the more easie access to each other, prov'd very fatal to me; for I could not come to this Gentleman, (for whom I had a very great kindness) to communicate my Secrets and desire his advice therein, but there would be present intelligence given to this Neece, or my Wife, where I was; that when I came thither as to a Friend, to disburthen my self of sorrow, and to receive his comfort, (having little at home) I was most commonly prevented.

Back doors unhappy.

And because to prove the truth of things, it may be required to instance, some Examples; this Dean having had some discourse

THE VINDICATION OF RICHARD ATKYNS

with my Wife, was (upon intelligence as aforesaid) sent after Nine of the Clock in a Winter Night, through a Sick Child's Chamber, of which Sickness she dyed, the only Child of her Parents; and conducted by my Wives Maid, who March'd with a Candle before him to the place where this Gentleman and I were together, refreshing our selves: We wondred to see the Dean there, at that time of Night, and to be brought through a private passage, that none but our selves should know, and never there before, nor knew he the Master of the House; but in short, he fell upon the business, with design, to make me promise to renew the said Lease to Trustees for her; which when he could not accomplish, I ask'd him whether I had ever promis'd him so to do, he answer'd I had not; and till this time my Wife reported to the Gentleman of this House, and several others, that I had so promised.

Not long after, the Dean came to Town again, and I desir'd to speak with him at his Lodging, but he alwaies put me off saying, he would come to me; and when he came to my House, (though he knew I was within, he would never speak with me, till he spake with my Wife first; and once he had the confidence to invite me up to my Wives Chamber in a Morning, by this Maid who conducted him to the Neighbouring House, (I not hearing that he was in the House before) which much amaz'd me; knowing how censorious my Wife had ever been of those that entertained men in their Chamber. *Great confidence.*

Yet some are of opinion, that God sees no Error, in his Saints; and the Authority of my Lady *Acheson* may make that a Virtue, which in another shall be judg'd to be Vitious: But if persons should put a reputation upon things, why should not the Zealous Lady *Norton*[148] be Cannoniz'd for frequenting of Sermons, and claping her Hands upon every defeat given to the King: Besides so grave a man as the Dean to visit her, must challenge as much priviledge as if a Nun had her Confessor come to her. When I came up to him, he fell upon the old Theam, how just and fit it was to renew this Lease to Friends in Trust for my Wife (though I told him of the Debt *A design frustrated.*

[111]

RICHARD ATKYNS ESQUIRE

upon it) and stroak'd his grey Beard in confirmation of the Conscienciousness of it; but for all his gravity, and the power he had upon me, I refus'd his proportion; at which he was very angry, and flew away in a great Passion: Afterwards my Creditor offer'd him the mony agreed between us; and Interest from the time it should have been paid; which he was so far from accepting of, that he treated with him to accept of the Money and Interest that was due to him of a Friend of his, and nam'd the person for his Friend, that to my knowledge (his Debts paid) is 20000 *l.* worse than nothing; which when my Creditor acquainted me with (apprehending that if he would take his mony, a Lease might be renewed against me) I thought it high time to make my applications to the Prebends, to acquaint them with the Design I supposed against me, and addrest my self to them, by the Bishop of that Diocess; who was so Noble, as to cause them to declare their Innocence and Integrity in the business, and their willingness to let me have it at the Fine agreed upon; But the Dean still persisted in his recusancy, and took himself to be Chancellor of Equity, as well as Dean of *Gloucester.*

16th Sigh.

—*Plead thou my cause, O Lord, with them that strive with me! and fight thou against them that fight against me; let them be confounded and put to shame that seek after my Soul; let them be turned back and brought to confusion, that imagine mischief for me; for they have privily laid their net to destroy me without a Cause; yea, even without a cause have they made a Pit for my Soul; my Lovers and my Neighbours stand looking upon my trouble, and my Kinsmen a far off: They also that reward Evil for Good are against me; but 'tis not an open Enemy hath done me this mischief for then I could have borne it, nor a known adversary, for then I could have avoided him; but it was even thou, the Wife of my Bosome: We took sweet Council together for many Years, and walked in the House of God as Friends; but as for me, I will call upon God, and implore his aid to deliver me but of this misery.*

The Devil fittest to Sow Tares.

But these contrivances were still only preparatory, for the Devil having begun so prosperously, as to separate so near relations, would be sure to strike home at last; knowing that *Corruptio optimi est pessima,*[149] and therefore adviseth a privy Council to carry on the Work so happily begun, least malice should fail of its end: She therefore resolves upon persons for her Council that have two marks upon them, the one to hate me, the other to love themselves; I mean advantage, which she is very apt to promise upon the least design: The first she pitches upon, was one old in injuries to me; the second, one whose Mouth watred at a Mannor of mine; the Third, one that came from far to govern her, and her Estate: Of all which, I shall give you a short Character, and begin with Sir *John Denham* his description of the first *viz.*[150]

He shuffles with his Legs, and rids no Ground; he sputters with his Tongue, and utters no sence; he speaks all Languages at once, as if he had been at the Confusion of *Babell;* he thrusts himself into every ones concern, and alwaies gives Council before he be ask'd: A Collonel (so call'd) that never drew Sword in Anger; a Statesman that never Writ true *English;* a man that whispers nothing in every mans Ear, and would be the true *Sir Positive at all,*[151] if he were Knighted: This is Chief Secretary of the Council. *The First.*

A Church-man, that suffers none to surfeit in his House; he is himself of the *Spanish* Diet (I hope not of their Faith) one Fig, and two Almonds will satisfie him at home, but abroad his Stomach is so extensive, that he can swallow a whole Mannor at a Bit; he is much addicted to the Scurvy, and can Eat and Drink a Dozen of Oranges[152] at a Meal, in a Physical way; he will rather break his Tenants backs, than their Bellies; and is so consciencious, that he will rather Prey upon them, then Pray for them: Towards Twelve of the Clock he is as humble, and nimble with all, as a mendicant Frier; but when he is in *Cathedra* as Majestical as the *Pope himself:* He is good at most things but Preaching, and better at separating than joyning together, though his Function teach otherwise; he is the Champion of Ladies, as far as fits his own turn, and practises the 3 Chapter of the 2 Epistle of Saint *Paul* to *Timothy* to the life: To conclude, Ten such Church-men in *England* would endanger *The Second.*

a Second Rebellion (blessed be God, I know but this one.) With these perfections he stands to be president of the Council, but the Ladies will not permit him, unless he Cuts off his long Beard.

The Third. A thing long, lean, and without a Beard, in habit much like a woman; but whether Man, Woman, or Hermaphrodite; Wife, Widdow, or Maid, I know not; she is all of a peece, and not at all like the Kings Daughter, *Glorious within;* she Cannot pronounce *Shibboleth* plain; yet for all the Church-mans Learning, this is thought most fit to be president of a Ladies Council, most of them being also Women: for the rest of her description, I leave you to Sir *William Davenants* Song of *So Old, So wondrous Old:*[153] And if that should fail, to Master *Randolph's* Verses upon such another Beauty.

> *Lower than Gamut sunk her Eyes,*
> *'Bove Ela though her Nose did rise.*

As for her Virtues, they are so Eminent, that they bear proportion with her Beauty.

The Devil president of wicked Councels. But though these contend to be President, yet really none of them are fit for the place; for in Councels wherein the design is to separate Man and Wife, there is none so proper to be President as the Prince of Darkness, who certainly may claim the most undoubted right; the rest of the Council being women seduced through mis-information, I shall not trouble my self with, hoping they will not trouble them-selves any further, when they know the truth of the Cause: All these conjoyn'd together against a single Person, obscur'd with great Debts (who made no preparation to defend himself, relying on his own Innocence) together with an Estate Real and Personal to back them; might (without the least Difficulty, as well as Honour) promise themselves success, and be as confident as *Thomas Aquinas* was, when he said,

> *Jam contra manichæos conclusum est.*[154]

THE VINDICATION OF RICHARD ATKYNS

There only wanted now an opportunity to set on Foot the Decrees of this Councel, which happened not long after; for a Creditor, who obtained an Execution against me, upon a House Debt, which my Wife ought to pay, threatned to arrest me; which to prevent, I sent a Gentleman I thought most in her favour, to desire him to forbear seizing on my Person or Estate, for a Month or Six Weeks, and he should not fail of his Mony; his Answer was, he would not stay at all; and if present Mony were not paid, he would take what Course he could to get it: I desired the Gentleman to acquaint my Wife therewith, which he did in my presence; I took that occasion, to tell her of the danger, the disgrace, and the loss that must necessarily follow, if this Debt were not paid; and beg'd of her (as if it had been for my life) that she would pay it, and I would allow it; and though she had then in her Custody, above 1000 *l.* worth of Jewels, and Plate, which she still unjustly detains, I could not prevail with her to do it, although the Debt were little more than 60 *l.* I thought this very hard, and yet was unwilling to seiz my own Goods, to pay it against her will; but told her, I would secure my Person, and some Pictures that were like to be undervalued, in case they were seiz'd upon: For the absence of my Person, she seem'd not to trouble her self much, but for the Pictures, she was in such a toss, that she utter'd such Language, as I never heard from any Lady, and told the Gentleman, in whose custody I put them, that he had plundred her House, *&c.* *Good councel rejected.*

These things made me examine in what safety I was in the House, Answer was made me, that there had been no Key to the Street Door for two Months, (for indeed, I was as a Stranger in the House for above a Year before) when I considered of this, as also, that my Wife exprest little trouble at my last Arrest for her own Debt, and that there was but one Servant I could trust; I resolv'd to take a Lodging, and presently to remove; which I did in that hast, that I left all things in disorder behind me; I often sent to my Wife, to know how she did, and to bring me things I had occasion to use, but I could not receive them without a note under my hand, and sometimes not at all; and the Messenger complain'd that he was alwaies dog'd from thence, and having but one to look to me, I was depriv'd many times of a Servant:. I then prepared for the Country to put my business in the best order I could there, and secure those Goods: And as I went to take Coach, I escaped Arresting very narrowly.

In the Country I stai'd but a short time, being alarm'd up to Town about a Charitable use, alledged to be given by my Wives Father, out of her Estate in the *Strand*, for the Payment of which, there was a Decree by the Commissioners of Charitable uses against me, for about 400 *l*. for the Arrears, and 10 *l*. per ann. for ever, which would have been also a leading case to the Payment of a greater Sum, by this as well as being lyable to my Wives Debts, Troubles and Suits, you may perceive I had all the Evils of the Estate, though no benefit thereby; and I heard my Wife took no care at all, to prevent the Confirmation of it in Chancery: Whereupon I went with what speed I could to *London* to our Council, to instruct and Fee him for the hearing, which after I had done, and was taking my leave of him, he was the first that told me of the sad News, of my Goods being seized upon, and that if I would stay, he thought I might see my Wife; with that I staid, and half an hour after she came; he said to her, that if she would walk into his inward Room, he would shew her that he hoped would please her; when she came in, I Saluted her, and she seem'd well enough contented to see me: Afterwards he made several propositions between us, whereof one was accepted, and others like to be agreed on; but upon conference with her Sweet Neece, all was contradicted again, and she did never since imploy the said Councellor, for endeavouring a Reconciliation between us.

Honesty mistaken.

Here I may not omit to tell you that there were several things reported of me Scandalous and false; as that I put this Creditor upon seizing the goods of the House (which is as much as to say, I contrived mine own dishonour and ruine) 'tis more likely to be my Wives contrivance, who hath built her whole design upon this

THE VINDICATION OF RICHARD ATKYNS

Foundation: For after this my Credit was wholly blasted, that my familiar Friends seem'd to shun me, the reason they gave for it was, because my Study-Doors were not broken open, and that some of my wearing Cloaths, (though in another mans Chamber) were preserv'd: 'Twas also reported that my Wife was surpriz'd, when she agreed as aforesaid; whereas I dare swear, the Councellor that advis'd it, did it from the integrity of his Heart for both our goods; and that he had as much or more kindness for her, than he had for me; nor could what was reported be of any advantage to him, or to me if true, but just contrary: 'Twas also reported, that I did not Visit her, nor acquaint her with my going out of Town; that I sold the Coach-Horses, which I confess I did, with her consent, and such trifles, which I should blush to mention, had they not come from her; and that I had left her, and not she me, by way of fallacy, because she lives in the House still; as if her promise in Marriage were made to the House, and not to my Person. To which I answer, that she might have prevented the mischief that parted us, if she had pleas'd, having warning enough; and that I durst not see her in that house, beseig'd with Bayliffs, for fear of being Arrested, which had then utterly undon me: I will not say she would have betray'd me her self, but I am sure those about her, which she will not part withal, would.

More Scandals.

However, when I came into the Country, I gave her an account where I was, but could receive no answer of many Letters sent to her from thence, nor never since, though I have written at least four or five Letters to her since I came to *London;* nor did she ever see the Servant by whom I sent them, but by accident: After this I met her at my Uncle *Baron Atkyns*[155] his House by chance; who had been alwaies exceeding kind to her in all her distresses, and was so in this; for he invited her to his House, and lodg'd her there for a fortnight at least, he was (as others were) possest of an ill opinion of me, and that I was a strange Monster of a Man; but when I had in some measure undeceived him, he began to be of another mind, and to examine the truth of things; which she perceiving, and that 'twas his opinion we should live together; she suddainly remov'd her Lodging, without acquainting him therewith: When we were there together (the good old Gentleman, having made his peace with God, desirous also to make peace in his own Family,) propos'd, that we should ſtay there and lie together, and agree the matters in difference between us, which I consented to, and she refus'd; he then propos'd, that the matters in difference might be referred to him, assuring her, she should be no looser by it; which she also refus'd: He also propos'd, that we might talk of business there; her answer was, she came not thither about business, and so she took leave and turn'd away, and went out with the same Gentleman she came in with, leaving me there; and lodg'd at his House without my knowledge or consent: this being matter of Fact, let the World Judge, whether I left her, or she me.

More good Councel rejected.

After this, I sent two Ladies of Quality to propose an accommodation to her, and desired a meeting; at leaſt, that I might know my fault, for which I was so severely Punished; from whom I could receive no other answer than this, that when they touch'd upon that ſtring, she fell off from that discourse to another; as if the very discourse of re-uniting us, had been infectious; not answering so well as *Felix* did to St. *Paul, When he reasoned of Righteousness, Temperance and Judgment to come: Go thy way* (said he) *for this time, when I have a convenient Season I will call for thee.* For she would not give them hopes of an Answer; but I truſt, he that caſt [the] *Dumb Devil out of the posseſt,* will in his good time reſtore her to her Speech even upon this Subject; for whoever is acquainted with my Lady Acheson, muſt needs know that she can ſpeak.

24 Acts v. 25.

Matthew 9. 32. 33.

More Scandals.

Then I made my addresses to a Lady of Honour, to whom she made her applications firſt; who was perswaded by her that I would not let her have a House to live in; that I had got her to Sell great part of her Eſtate to pay my Debts, and would have her Sell more; that I was vaſtly in Debt, and would not give her an account how much: That I was a great Swearer, a Drunkard, &c. and made my Case so deſperate, that this Lady (though a very prudent person, and my friend) would hardly hear me; but said, the best way would be to turn my self over to the *Kings-Bench*, and Compound my Debts, and let my Wife have what was left to keep herself, and me in Prison. After I had leave to speak, and had in some measure undeceiv'd her, declaring how hard it was to judge upon hearing one side only; She said she had proposed to my Wife to have the business in difference between us referred to two Lawyers indifferently chosen to determine, and nam'd the persons; to which I readily contented (though the person nam'd for me, was of my Wives Councel as well as mine, th' other wholly a stranger to me) I humbly desired the Lady to put it on, and give me a timely answer; which was, that over Night my Wife agreed to it. But contradicted it next Morning; thinking (as I suppose) that Lawyers could not give her more than she had already: But she said my Wife desired a list of my Debts, which she thought, might bring on a Treaty: In two or three daies, I sent it by this Noble Lady, with two propositions under my hand, to this effect, *viz.* First, That both our Debts, and a particular of both our Estates, may be referred to two Lawyers as aforesaid, with power to pay the said Debts, out of the said Estates and to settle the remainder according to Justice and Equity: Otherwaies, Secondly, That I may have what is Lawfully mine, out of which I will pay my Debts, and discharge her from them, so she will do the same to me: Which though she delivered with her own hand, yet had she no better answer than this; That if I would send to her four, or five Daies after, she would give me an answer: So the Lady return'd the Papers, as hopeless to do any good; and after six Daies, and twice attendance upon her, and another Letter from me; she said, her Counsel advised her to give no Answer.

—*Lord, how are they increased that trouble me! many are they that rise up against me, and say, there is no help for him in his God: Thou hast put mine acquaintance and Friends far from me, and made me to be abhorred of them; but thou O Lord art my Defender! and canst lift me up again, for thou wilt save the People that are in adversity, and bring down the high*   *17th Sigh.*

looks of the Proud: And though false witnesses rise up, and lay to my charge things I know not, and reward me evil for good, to my great discomfort; yet let them not, that are mine Enemies, Triumph over me, nor be beleived when they speak Scandalously against me; but let the mischief of their own Lips fall upon them.

Having spent the greatest part of the Summer, in attempting a reconciliation with my Wife, in which I failed; about the middle of *August* I retir'd into the Country to Baron *Atkins* his House, after many kind, and importunate invitations thither; who us'd me as his own Son in all things but this, that he gave me more outward Civility and Respect: He seem'd much concern'd for my Wives unkindness to me; and though he was troubled with very great pains, and drew near his end (being in the 83 Year of his Age)[156] he spake very obligingly to me; and his Lady, Son, and Daughter in Law, were more than ordinary Civil to me; the Servants also were so Well instructed, that, though I had a free Command of the House, I need not have ask'd for any thing, it being brought to me before hand; so that for the freedome of it, I might have taken it for mine own House, in all other respects save that I was us'd with much more kindness and respect, and fed at a more plentiful Table: For the Baron was never so well pleas'd, as when he had the best Meat the Country could afford, and his Friends and Neighbours to eat of it, (for he eat little himself:) The discourse at his Table, was much after the description of St. *Augustines,* but his Meat better than *Augustines* for an *English* Stomach. To come to the matter in hand; the discourse of the Married Ladies was chiefly, whether Love or Duty, obliged them most to their Husbands; acknowledging both to be ties upon them: There was publique Prayers constantly, four times a day; and seldome a Servant missing, whose business was not a just excuse; so that this was not in Name, but in Deed a Religious House.

After I had been there about ten daies, I sent my servant to *London;* of which I thought fit to give the Baron and his Lady notice,

Baron Atkins his kindness to me.

The Order of the House.

who upon discourse together, agreed to invite my Wife thither, but first ask'd whether I approv'd of it, to which I gave my full assent; so my Lady did me the favour to shew me the Letter, which was a very kind invitation, owning that I was there with them; I gave my Servant order to deliver it as soon as he came to Town, and into her own hand: His answer was, that he was twice at her Lodging, and could not be admitted to speak with her, though he sent up his message; so he was forc'd to send up the Letter by her Servant, who said, her Lady had appointed her to receive it; and though I staid there, or there abouts six Weeks after, the Lady receiv'd no answer at all of her Letter; God grant it was not to avoid good Example, for there were as good Wives as ever I saw; but my Wife continued all the Summer in *London*, though she never staid a whole Summer there for 30 Years together, except one, if that? for had my occasions been never so great and urgent, even to my ruine, I had not a quiet day after *May*, till I carried her into the Country; complaining she wanted Air, though she knew, I had not many times wherewithal to do it: but what Love cannot do, Revenge may; and what neither of them can do single, Love and Revenge may do together; I mean her Love to that company she forsook me for, and her Revenge upon me; whom she so solemnly promis'd before God and man, never to forsake, in these words, *viz. For Better for Worse, for Richer for Poorer, in Sickness and in Health, to Love, Cherish, and to obey till death us depart, &c.* Now several Questions may arise upon these words.

His kindness to my Wife.

1. Whether a Husband being reduced in his Estate, and his Wife giving out that he is a Scandalous liver, without proof, can nullifie this promise?
2. If they be prov'd, it can dissolve this knot, unless they both agree to the dissolution?
3. If this be agreed, whether it may be done, since the promise is, *Till death us depart*?
4. Who can do it?

4 Questions.

RICHARD ATKYNS ESQUIRE

The Questions answered.

In answering which Questions, I shall not irreverently look into the Ark, nor meddle with the Priests Office, which belongs not to me; but because this is the most essential point of the whole Treatise, and wherein I am most concern'd, I shall not deliver mine own opinion; but shall give the opinion of a Sober Divine, *viz.* To the first Question he saith.

To the first Question.

That neither decay of Estate, nor scandal, nor any other thing without proof can nullify this promise, or be any just cause of Separation; for *De non apparentibus & non existentibus eadem est ratio.*[157]

The second he saith, hath two parts: First, If it be prov'd it can dissolve the Knot: Secondly, whether there can be a dissolution of it unless both parties agree.

To the first he Answers two things; First, That decay of Estate if prov'd, is not sufficient to dissolve the Knot, because the Divine Law allowes of no such cause; and this is expresly provided against in the words of Marriage, *for Richer for Poorer.*

Secondly, Nor is every Scandal sufficient if it were prov'd: The Divine Law allows only of one cause as sufficient, viz. *Adultery*, Mat. 5.

To the second.

To the second part of the Question he sayes, first, If the Adultery be on the Wives part, the Husband may certainly without the Wives consent, dissolve the Marriage by the Divine Law. Secondly, If the Adultery be on the Husbands part, the case is not altogether so clear concerning the Wife, that she may put away her Husband. Among the *Jewes,* it was in no case permitted to the Wife to put away her Husband; and the *Greek* Fathers (as *Grotius* upon *Matt.* 5. tells us.) That Anciently among the *Christians,* it was not allowed to the woman to dissolve the Bond of Marriage, no not for Adultery; But our Saviour seems to allow it on both sides, *Mark* 10, 11, 12. and St. *Paul* 1 *Cor.* 7, 10, 11.

To the 3d Question.

To the third he saith, that notwithstanding the words of the promise (*till death us depart*) the Marriage may be dissolved for the cause of Adultery, and this though both parties be not agreed: He grants indeed, that neither party is bound to make a divorce, no not in case of Adultery; because the Law of Divorce is only permissive: He likewise grants, that the words of Marriage might be so fram'd, as to oblige both parties beyond what the bare nature of Marriage doth: So two Persons when they Marry, may solemnly oblige themselves never to be voluntarily absent for a day from one another, which Marriage doth not tye them to; so likewise they might mutually oblige themselves not to make use of the permission of Divorce in any case, but doth not think that these words (*till death us depart*) are of that force, or that they were so intended: First, because among the *Jewes,* who had the greatest liberty of Divorce, the like expressions are us'd, as if nothing but death could dissolve this knot; and yet this did not take away the liberty of Divorce St. *Paul* tells us, *Rom.* 7.2. *That the woman is bound by the Law to her Husband as long as he liveth,* and yet in case of Divorce, the woman was free from her Husband even while he was alive: Secondly, Constant practise among our selves, hath so interpreted these words, as to be no prejudice to Divorce.

To the Fourth he saith, That by the *Christian* Law either party may do it, but by the Lawes of every particular Nation those Courts and Judges that are appointed by Law, and in *England* it is to be done by Ecclesiastical Courts only. *To the 4th Question.*

To add to this the opinion of Bishop *Hall*[10] (who is generally the Ladies Doctor, and particularly my Wives) in his Cases of Conscience upon Divorce, he concludes, that a man cannot put away his Wife for any Cause, except Fornication, or Adultery, no not for Unbeleif, or Heresie; and after saies these words; Therefore to separate from Board, or Bed, is no better than a presumptuous violence. It is the peremptory Charge of Christ, *what God hath joyned together let no man put a sunder, Matt.* 19. 6. In all lawfull Marriages 'tis God that joyns the Hands and Hearts of the Married: How dare man then undo the work of God, upon devices of his own? what an impious sauciness is it to disjoyn those whom God hath United. *Bishop* Hall *his opinion.*

RICHARD ATKYNS ESQUIRE

But neither these, nor any other I ever heard of, hold it lawful for a Woman to separate from her Husband without any proof at all, which is my Case. If Actions be the true Interpreters of Thoughts, 'tis more than probable some Wives think, that if they be actually chaſt in their Bodies, their Husbands are so much in their Debt, that they may abuse them all other waies; wherefore I shall the less judge such as buy Milk by the Quart, when a Mans own Cow shall give a good Soop of Milk, and kick it down when she hath done.

The Kings Farriers Case.

'Tis a new fashion but I hope will never take, for a Wife to Seise her Husbands Goods, and then to complain and separate from him, to whine and bite.

New lights.

Yet the new lights of these times may make Ladies believe they can see further than those of old, because Age is blind holding forth that the Yoke of Marriage is too heavy for a tender Lady to bear, who is more of Spirit than Flesh; they have a Salve for every Sore: Sometimes breaking a Heart into three peeces, and then taking five peeces for a word of Consolation, and so according to the occasion: But a Divorce ought to bear the greateſt price of all, since there are so few, (if any,) can do it; for what would not one give to enjoy the thing belov'd, and to be rid of the thing deteſted; besides in reason, for a Woman of Eſtate, to endure a Husband of 29 Years ſtanding, is as great a misery, as for a Father to live till his Heir be 50 Years of Age: 'Tis civil for them both to depart sooner. Wherefore this may be called the Old Law, which by the entrance of these new Lights is abrogated.

Whilſt she was in *London* she found her self a full and more pleasant Imployment than taking the Country Ayr, in Writing and sending to *Glouceſter* to hinder my renewing the Lease aforesaid; but the Chapter were so far engag'd to the Mortgagee, that both of us miſt of it; then she made in to two ingenious Lawyers, fit for her turn, who Travel'd into *Ireland* to learn honeſty, with whom I had intruſted some writings of an Eſtate not made over to her, and she prevail'd with them to keep both writings and Rent from me, to make all things so litigious as that I might not get any thing for a Lively-hood, but be forced to a Complyance with her upon Necessity; and so she played the Tenants games for them, who were glad of any occasion to keep their Rents in their hands; that instead of receiving any Mony, I was compelled to spend Mony in Suits: Upon which one was so much encouraged, as to attempt to pay me with ill Words and Blows, though I cannot believe she sent him; nor can I hear he hath brag'd of his payment since; and though I have been informed of all hands, that the breach (of I know not what) is past reconciliation; yet I sent my Servant since my coming to Town, to see her, who (as he informed me) could neither see her, nor any of her Servants; as also a Gentleman who formerly hath been admitted, but coming from me was rejected: She is so angry that her designs are frustrate, and that *Mordicai* will not bow lower. This Spiritual pride which transform'd *Lucifer* from an Angel of Light into Darkness, hath taught her first to despise, and then to hate, because despis'd; nor will I wholly excuse my self from imperfections, which may give her some cause of offence, though they have been much heightned by those that get their living by it; 'twould be a great disappointment to them that faults on both parts should be laid aside, and we joyn together to deceive the deceivers, which is no deceipt; for if we forgive not each other, how shall we both expect forgiveness from God: But when she sees her errour, and returns, I shall be ready to do my part and meet her; in the mean time, while she flourishes with what is mine, 'twill be natural for me, to keep my self from that which breaks down Stone-walls; and I hope by the Sun-shine of Gods Grace to break through this Cloud in time, not as *Hercules Furens*, but *Christianus Militans*. *The danger of Pride.*

Not to be further tedious, I shall close up all, as *Homer* did his *Iliads* in a Nut-shell, and leave it to the Readers judgment: My Education and Actions have been alwaies free and without any ill design upon others, I have rather practiz'd upon my self; my nature so flexible as to receive the impressions of others designes upon me, my trusts so confident (thinking others as honest as my self) *The Sum of all.*

RICHARD ATKYNS ESQUIRE

that they have been the cause of my misfortunes; my want of diligence and suspect of others, have made me alwaies play the after Game; and nothing but sence of Honour did ever give me diligence enough to prevail against my Adversaries; *Gods providence* hath protected me in all things, and I have deserv'd it in nothing; I have had great tryals upon me, but never any like this; Of which I may say as *Julius Cæsar* did, after his last Battail with Young *Pompey*, *That in all other Battails he Fought for his Honour, but in this, for his Life.* So may I say, all other tryals were fleabitings to this, and the *formidablest Army* I ever look'd upon of less Terrour than this of *Gnats*, as if there were a second *Pucell of Orleans*[158] in the World; yet *Solomon* and *Sampson* are my Companions, for those that neither Wisdome, nor Strength could Conquer, have been subdued by Women. They were first form'd out of Mans Rib, and made Bone of his Bones, and Flesh of his Flesh; so that that they may if they please, the more familiarly smite him (as *Joab* did *Amasa*) *in the fifth Rib.* I should be laught at to advise others, who could not advise my self; yet this I must needs say upon the whole, that Government in Women Covert, (for I speak of no other) is most commonly distructive, even to themselves, for it often gives occasion for a third person to deceive them both; Nor do I think the *Empire of Rome* best Govern'd, when a two year old Child rul'd the *Empress,* and she the *Emperour:* Yet where men are Beasts (which I hope is not in this Case) I have known several discreet Women that have preserved their families from ruine, and wish the Lady I Write of, may be so prudent as to do the same.

18th Sigh. —*Lord remove all misunderstandings, and Animosities between my Wife and me! and those from both of us, which are, or may be the Cause of them; and bring those to condigne shame and punishment, which make it their business to divide whom thou hast joyned together. Forsake me not O God in mine old Age! when I am gray Headed, that thy Strength and Power may appear in my weakness; though they Curse, yet Bless thou, and let them be confounded and Clothed with shame, that rise up against me; but* let thy Servant rejoyce, that I may say with David, 'tis good for me that I have been in trouble, that I may learn thy Statutes; Lord grant that they which Sow in Tears may Reap in Joy, that as I have been like to Job in my afflictions I may also be like him in his patience and his end.

FINIS.

Notes

1. This engraving by David Loggan depicts Charles II seated on the throne. On his right is Archbishop Sheldon, on his left, Edward Hyde, First Earl of Clarendon. Above the king are the words 'per me reges regnant,' (through me do kings reign), expressing the divine right of kings, and 'iustitia stabilitur solum' (by justice is the throne maintained). Below is General George Monck, 1st Duke of Albemarle, who fought first for the Royalists and then the Parliamentarians. He was instrumental in the restoration of the monarchy. 'Cedant arma togae' (let arms yield to the toga) can be rendered as 'let the military relinquish control to the civil powers'.
2. Haarlem long laid claim to be the birthplace of printing, but this claim relied on a story that it was invented by one Laurens Janszoon Coster. Atkyns has conflated the Gutenberg and Coster legends.
3. Based on calculations at measuringworth.com.
4. Hills seems to have adapted his allegiances effortlessly, printing the masterpiece of royalist propaganda, the Εἰκὼν Βασιλική [Eikōn Basilikē], anonymously in 1649 before becoming Printer, first to the Anabaptist Congregation and then the Commonwealth. At the Restoration he assumed an Anglican character and was appointed Printer to the King. In the wake of the Popish Plot, he expended much energy into hunting out Catholics in the printing industry, then realised that this might alienate him to the King and converted to Catholocism himself.
5. That is, the Court of the Star Chamber, which was abolished by Parliament in 1641 (the act of 17 Car. mentioned by Atkyns below). The use of this and other conciliar courts to circumvent the courts of common law was a great bone of contention between parliament and the monarchy.
6. The Grand Remonstrance, presented to Charles I by Parliament on 1 December 1641 and published on 15 December, listed its grievances against the king.
7. Robert Devereux, 3rd Earl of Essex, was captain general of the parliamentary forces at the beginning of the Civil War. On raising an army, members of Parliament declared that they would live and die with the Earl.
8. Parliament entered into a Solemn League and Covenant for Reformation and Defence of Religion in order to enlist the support of Scottish presbyterians in their fight against the royalist forces.

NOTES

9. The Statute of Monopolies, passed by parliament in 1624, was intended to curtail the issuing of royal patents such as that for printing law books which Atkyns sought to uphold. By only permitting patents for original inventions, it ushered in the modern concept of patenting.
10. Joseph Hall (1574–1656), Bishop of Exter (from 1627) and then Norwich (from 1641).
11. Atkyns is referring to Catholics.
12. John Stow's *Survey of London*, first published in 1598.
13. Johannes Gutenberg of Mainz, of course. Gutenberg was not a knight but Atkyns clearly believed that only a man of high social standing could have invented something as important as printing. This is the same sort of silliness which leads people to claim that Sir Francis Bacon wrote Shakespeare's plays, and which caused Robert Greene to describe Shakespeare as 'an upstart crow beautified with our feathers'.
14. Sir Richard Baker's property in Oxfordshire was seized to pay the debts of his wife's family. He wrote *Chronicle of the Kings of England* (1643) while in the Fleet Prison for debt and he died there in 1645.
15. Like Baker, James Howell (1594–1666) was imprisoned in the Fleet for debt. His *Londinopolis. An Historical Discourse; or, Perlustration of the City of London and Westminster* was published in 1657.
16. John Islip (1464–1532).
17. This book, an edition of Rufinus's *Expositio Sancti Ieronimi in symbolum apostolorum*, does, indeed, bear the date M.cccc.lxxviij . xvij. die decembris (17 December 1468), but it has long been recognised that this was a mistake for M.cccc.lxxviij (1478). The printer was formerly believed to be Theoderic Rood but this attribution is no longer accepted.
18. Lambeth Palace, London home of the Archbishops of Canterbury.
19. Atkyns is conflating Gutenberg's claim to have invented printing from moveable metal type with the rival claim of Laurens Coster in Haarlem. Actually, moveable metal type had been used in Korea since the early thirteenth century.
20. William (not Robert) Turner, Master of the Robes.
21. Frederick Corsellis seems to have been a figment of Atkyns's imagination.
22. There was a short-lived press at St Albans in the late fifteenth century, associated with the abbey there. There is no known royal connection.
23. The heads of the Court of Common Pleas and of the Court of the King's Bench were both styled Lord Chief Justice.

NOTES

24. The Stationers' Company was incorporated by royal charter in 1557, although it had been formed in 1403.

25. The letters patent granted to the Company by James I in 1603 specifically refer to the 'mystery or art of stationers' and a mystery or mistery in this context means a trade or craft, not anything mysterious.

26. Richard Tottel (d. 1594) was granted a patent by the King for publishing books of common law in 1557, the year in which the Worshipful Company of Stationers was founded. He was a prominent member of the Company and served as its Master in 1578 and 1584. There tended to be a certain conflict of interest between the individual patentees and the wider interests of the Company.

27. Nicasius Yestweirt inherited Tottel's patent. He was a Clerk of the Signet, a rather pointless bureaucrat who acted as an intermediary between the Patent Office and the Secretary of State. No doubt this position gave him an insight into the value of such a patent.

28. Thomas Wight and Bonham Norton were granted the patent to publish common law books in 1597 until it expired in 1629.

29. John More was granted the patent in 1629 and swiftly sublicensed others to print law books.

30. This nonsense about ostriches burying their heads in the sand dates back to Pliny the Elder, and possibly much further.

31. The usual expression is *aliquis in omnibus, nullus in singularis,* somebody in general, nobody in particular (i.e. jack of all trades, master of none). Atkyns has *someone* in particular – he is not about to refer to the late Charles I as a nobody.

32. Sir James Palmer (d. 1657) was an amateur artist who helped Charles I create his picture collection. He was a governor of the royal tapestry works at Mortlake and, from 1645, served as Chancellor of the Order of the Garter.

33. Ferdinando Pulton's *An Abstract of All the Penall Statutes Which Be Generall, in Force and Use,* which was originally published by Richard Tottel.

34. Miles Flesher was one of those licensed by More.

35. John Bill and Bonham Norton paid Robert Barker (who had been Printer to Elizabeth I and then James I and in this capacity held the privilege to print the Bible and Book of Common Prayer in English) £5,000 to enter into partnership with his son Christopher in 1615. In the same year Christopher married Norton's daughter Sarah. Norton was, from 1613, the King's printer in Latin and Greek and Bill had a disputed claim to be the King's Printer in English (see Queen Mary, University of London, 'A Brief History of the King's Printing House').

NOTES

36. Perhaps a reference to the widely-held belief that a Roman temple to Diana existed where St Paul's Cathedral now stands, close to Stationers' Hall. Sir Christopher Wren found what may have been its ruins when digging the foundations for his new cathedral.
37. An apparent good and a completely genuine good.
38. Even Hercules [could not contend] against two. Atkyns has *ne* rather than *nec*, clearly a typo.
39. Junius Annaeus Gallio, son of Seneca the Elder, dismissed charges brought against Paul: 'And when Paul was now about to open his mouth, Gallio said unto the Jews, If it were a matter of wrong or wicked lewdness, O ye Jews, reason would that I should bear with you: But if it be a question of words and names, and of your law, look ye to it; for I will be no judge of such matters.' (Acts 18:14–15).
40. Atkyns is reminding Charles II of his military service on behalf of the royalist cause. His exploits are discussed in detail in the *Vindication*.
41. Poverty alone makes men thieves.
42. Sadly, this is believed to be an accurate description of Spartan society.
43. No doubt Atkyns chose this title deliberately to curry favour with the King. Advocating the assassination of Oliver Cromwell, it was written by Silius Titus and Edward Sexby under the pseudonym William Allen and published in the Netherlands during the Protectorate.
44. Queenborough, Isle of Sheppey, Kent, was an important wool port with imposing castle which was strengthened during the time of the Armada. Elizabeth came to Queenborough a number of times and there are several equally tedious anecdotes about her visits.
45. The hundreds of Dudston and King's Barton (Barton Regis) were taken from Gloucester and the city's walls were demolished after the Restoration in punishment for shutting its gates against Charles I. It is surprising that Atkyns should criticise this sanction which must surely have met with royal approval.
46. John Twyn was convicted of High Treason in 1663/4 under a statute of Edward III. *A Treatise of the Execution of Justice,* attributed to Twyn, was the mirror image of *Killing No Murder,* as it exhorted the people to put Charles II to death.
47. 'We, considering that seditious and heretical books, verses, etc. are daily published and printed by divers scandalous, malicious, schismatical and heretical persons, not only urging our subjects and lieges to sedition and disobedience against us, the Crown and our dignity, but truly also [urging them to] very great and detestable heresies against the Faith, etc. And wishing to provide in this matter, an appropriate rem-

NOTES

edy, of our special grace, etc.' Atkyns has omitted the reference in the original charter to 'the sound catholic doctrine of the mother church'.

48. Roger L'Estrange, *Considerations and Proposals,* 1663.

49. Charles II, addressing both houses of Parliament on 16 March 1664, complained, 'You gave me the Excise, which all people abroad believe to be the most insensible Imposition that can be laid upon a People: What Conspiracies and Combinations are enter'd into against it by the Brewers, who I am sure bear not that Burden themselves, to bring that Revenue to nothing, you may hear in Westminster-Hall.'

50. From reasonable, everyday cause/reason.

51. In customs it is soundness of reason not lapse of time which should be considered.

52. The Dutch jurist Hugo de Groot's *On the Law of War and Peace* was published in 1625.

53. Not all monopolies are at odds with natutral law, for they may sometimes be allowed by the highest authority in a just cause.

54. When the reason for a law ceases, the law ceases.

55. The King et cetera, by virtue of his royal dignity, is constrained on all sides to provide for the wellbeing of the realm.

56. Robert Gentilis's English translation of Paolo Servita's *Historia della Sacra Inquisitione* was first published in 1639.

57. This statute forbade the import of bound books for resale on pain of confiscation and a fine of 6s 8d. It also banned the sale of books 'at too high and unreasonable prices'.

58. In the original, 'It had…' begins a new sentence.

59. Within the King's treasury.

60. *Omnia iura in scrinio pectoris,* [the emperor carries] all laws in the shrine of his breast, was a dictum of Roman law which was later applied to the Pope and to monarchs. Fortescue quoted this in his *De Laudibus Legum Angliae,* first published in about 1543.

61. Edward Darcy Esquire v. Thomas Allin of London, Haberdasher, 74 ER 1131. This case of 1599 determined that a sole licence to import and sell playing cards granted by Queen Elizabeth to Darcy was invalid. The grounds were various but the court made clear that it disapproved of monopolies in general.

62. That is Thomas More's *Utopia.*

63. Which is to say, the book's type is defined by its principal subject. Atkyns has already expressed this idea in English and, as elsewhere in this pamphlet, the Latin merely adds veneer of legal authority to his arguments.

NOTES

64. This Richard Atkyns died in 1610. Tuffley is now a suburb of Gloucester.
65. Sir Edwin Sandys (c.1564–1608) of Latimer, Buckinghamshire. Atkyns describes his grandfather as a banneret but some later sources call him a baronet, which is quite wrong. Unlike baronetcies, the title of knight banneret or simply banneret, which ranks higher than a knight bachelor, is not hereditary.
66. Atkyns appears to be in some confusion here (not helped by repeated intermarrying in the Sandys family and a preponderance of Sir Edwins). William Sandys (c.1472–1540), 1st Baron Sandys of the Vyne, married Margaret, daughter of Sir John Bray. Margaret was also heir to her uncle, Sir Reginald Bray, Banneret, whose son, Edmund, was the first Baron Bray. Atkyns's maternal grandparents were Sir Edwin Sandys and Elizabeth Sandys, daughter of William, 3rd Baron Sandys (Kippis, p. 324) and Catherine Brydges, daughter of the 2nd Baron Chandos.
67. The famous and highly-regarded Crypt School.
68. He was a gentleman commoner at Balliol.
69. A zegadine or zegedine was a silver drinking cup. Atkyns was not unusual in dedicating his student years to the intemperate consumption of alcohol.
70. An impressive house of Tudor origins, the Vyne, near Basingstoke, is now owned by the National Trust.
71. Thomas Arundell, first Baron Arundell of Wardour.
72. Venter means 'womb', i.e. his second wife, Anne Philipson, whose mother was a Sandys. The Arundell genealogy is very confusing. The first Baron had a son by his first marriage, Thomas (later to be the second Baron), from whom he was estranged. Perhaps to spite him, he also called the son whom Anne bore him in 1616 Thomas. It would seem to be this son whom Atkyns accompanied in the early 1630s.
73. Arundell was made a count of the Holy Roman Empire for fighting the Ottoman Turks on behalf of Rudolf II but Queen Elizabeth disapproved of her subjects accepting foreign titles and he was briefly imprisoned. He was created 1st Baron Arundell of Wardour in 1605.
74. John Jewel, Bishop of Salisbury.
75. Calais.
76. William Russell, 5th Earl, later 1st Duke, of Bedford.
77. The English seminary at Douai.
78. Matthew Kellison (c.1560–1642).
79. Henri, 2nd Duc de Montmorency, joined forces with Louis XIII's brother, Gaston, Duc d'Orléans, in a rebellion against the King. He was executed at Toulouse on 30 October 1632.

NOTES

80. Since Charles I's mother, Anne of Denmark, died when Atkyns was 4, he must be referring to Henrietta Maria's mother, Marie de' Medici, Queen Dowager of France, who arrived in London 31 October 1638. She left England in August 1641. A new timber building was constructed next to the Banqueting Hall, 'possibly to impress Marie de' Medici' (Richardson, p. 127).
81. Lady Martha Acheson, daughter of William More, niece of John More and widow of Sir Patrick Acheson who died in 1638.
82. Colnbrook near Slough.
83. Convincing evidence of Atkyns's extravagance.
84. Quinsy, a serious throat infection.
85. Short of breath.
86. Wife of Sir John Finnett, Master of Ceremonies.
87. Sir Robert Pye, 1585–1662, was appointed Remembrancer of the Exchequer in 1616.
88. Thomas Wentworth, 1st Earl of Strafford served very ably as Lord Deputy of Ireland from 1632–9. It is presumably in that role that he allowed a lien against the Acheson property.
89. The strong Parliamentary garrison at Gloucester made it difficult for the Royalists to access the iron ore in the Forest of Dean to the West and the substantial wealth of the local wool trade. Whether Gloucester or the surrounding area was really as pro-Parliamentary as Atkyns suggests may be open to doubt. The merchants of Gloucester are said to have attempted to pay Royalist forces to leave them alone [Royle, p. 226], suggesting a pragmatic approach to loyalties.
90. Strafford's staunch support of Charles I made him many enemies in the Commons. After a failed attempt to impeach him, Parliament passed a Bill of Attainder, condemning him to death without trial. The King was faced with the choice of signing the death warrant or risking his throne, and possibly civil war. On Strafford's urging, the King signed rather than risk widespread bloodshed. Sadly, Strafford's sacrifice did little to retard the civil war or the King's own execution. the Εἰκὼν Βασιλική [Eikōn Basilikē], claimed by some to have been written by Charles I, says that the king's greatest mistake was to give in to Parliament's demand for Strafford's execution.
91. The Battle of Edge Hill on 23 October 1642 was the first pitched battle of the war. It was indecisive.
92. George Brydges (1620–1665), 6th Baron Chandos, a kinsman of Atkyns.
93. Son of Frederick V, Elector Palatinate, and nephew of Charles I.

NOTES

94. William Villiers (1614–1643), 2nd Viscount Grandison of Limerick. He died of wounds received at Bristol on 30 September 1643.
95. Thomas Sheldon.
96. Ripple Field was a notable victory for Prince Maurice's forces against Waller's, the parliamentary side losing at least eighty men.
97. Robert Devereux (1591–1646), 3rd Earl of Essex.
98. Caversham Park on the outskirts of Reading.
99. Elder brother of Prince Maurice.
100. William Legge 1608–1670.
101. 'But putting forth his horse, what with the falling of the staff too low before the legs of the horse, and the coming upon Dametas, who was then scrambling up, the horse fell over and over, and lay upon Clinias.' Sidney, *The Countess of Pembroke's Arcadia*.
102. William Seymour, 1st Marquess of Hertford.
103. Sir William Waller, a highly-regarded general on the Parliamentary side.
104. Robert Dormer, 1st Earl of Carnarvon.
105. Crookhorn near Portsmouth.
106. Ralph Hopton, 1st Baron Hopton. Despite being on opposite sides, he and Sir William Waller were close friends.
107. Taunton Deane, Somerset.
108. Chewton Mendip in Somerset.
109. Battlefield.
110. Gold coins larger than guineas.
111. William Seymour (1587–1660), 2nd Duke of Somerset, 1st Marquess of Hertford.
112. In southern Gloucestershire, eight miles from Bath.
113. Robert Burrell.
114. Richard Arrundell. He was not created 1st Baron Arrundell of Trerice until after the Restoration.
115. Sir Robert Walsh, knighted by Charles I at Edgehill, having recovered the colours of the King's Lifeguard.
116. Sir Bevil Grenville (1596–1643), MP for Launceston and staunch Royalist.
117. Richard Neville (1615–1676).
118. Ludovic Lindsay, 16th Earl of Crawford (1600–1652), a Scottish peer.
119. Colonel Sir James Long Bt (1616–1692), a Fellow of the Royal Society.
120. Lambourn, Berkshire.
121. Henry Wilmot, 1st Earl of Rochester (1612–1658).

NOTES

122. Thanks in part to Heselrige's Cuirassiers (the so-called 'Lobsters') blocking his own side's line of cannon fire, Roundway Down saw the complete destruction of the Western Association Army under Sir William Waller.
123. It was Charles I's attempt to arrest Heselrige (or Haselrigge) and four other members of the Commons which helped trigger the Civil War.
124. Tuck – a rapier.
125. 'I wrote these verses; another has stolen the honours.' – Virgil.
126. Sir Henry Wroth (1604/5–1661), royalist army officer.
127. A small, sturdy horse of North African origin.
128. Actually Sir Nicholas Slanning (1606–1643), MP for Penryn. He voted against the Bill of Attainder against Strafford.
129. John Trevanion, MP for Grampound in the Short Parliament and for Lostwithiel in the Long.
130. William Cavendish (1592–1676), created Marquess of Newcastle-upon-Tyne by James I and Duke by Charles II.
131. The siege of Gloucester was lifted when the royalist forces left on 5 September 1643 to intercept Essex's army at the first battle of Newbury.
132. Jacob Astley (1579–1652), later 1st Baron Astley of Reading, a Royalist army officer.
133. Cademan was physician to the Royalist army.
134. The Treaty of Uxbridge in early 1645 was an abortive attempt to end the First English Civil War. As well as a long list of parliamentary demands, the negotiations suffered from the fact that both sides believed they were in a position to win the war.
135. That is, divine law dictates that the Church is governed by a body of elected elders (rather than bishops). One of the conditions insisted upon by the Parliamentary side was the imposition of presbyterianism in England.
136. A composition (or compounding) is an agreement between a debtor and two or more creditors to pay a proportion of the debt, each creditor receiving the same proportion of his claim. Although Atkyns is describing proceedings for a civil debt, many faced sequestration of their estates at this time simply for being royalists. As has been remarked (Hill, p. 161), 'Sequestration became the means by which money income and personal estate, forcibly confiscated from the Royalists, went to finance the force that expropriated them.'
137. Parliament had set up a Committee for Compounding at Goldsmiths' Hall in 1643.

138. A small prison in Cheapside, named after one of the main commodities of the area.

139. A prison, often used for bankrupts, situated in Southwark. It was named after the Court of King's Bench.

140. William Lenthall, a Parliamentary supporter, was made Master of the Rolls in November 1643.

141. Hicks Hall in Clerkenwell was the sessions house for Middlesex.

142. John Barkstead, one of the regicides, was appointed Governor of the Tower of London in 1652.

143. Colonel Henry Norwood (1615–1689), Treasurer of Virginia from 1661 to 1673 and Lieutenant Governor of Tangier.

144. Elizabeth Boyle (*née* Feilding, d. 1667), was only created Countess of Guilford in 1660.

145. Sir Robert Viner or Vyner (1st baronet, 1631–1688), a wealthy goldsmith, was Lord Mayor of London in 1653–1654. He lent vast sums of money to Charles II.

146. That is, he would first take counsel. – 'Then she spake, saying, They were wont to speak in old time, saying, They shall surely ask counsel at Abel: and so they ended the matter.' 2 Samuel 20:18.

147. A rumper was a member or supporter of the Rump Parliament, the parliament which had been purged in 1648 of all those opposed to trying the King for high treason. Since the members of this Parliament, completely devoid of democratic legitimacy, were responsible for the execution of Charles I, the term 'rumper' was extremely pejorative in Royalist circles. However, it seems unlikely that his wife's advisors were literally associated with the Rump Parliament. Rather, Atkyns is simply using the strongest form of abuse he can think of, or perhaps that which he thinks will evince the most sympathy from the King.

148. Presumably Frances Norton (1644–1731), a writer of religious verse and prose.

149. The corruption of the best is the worst.

150. Sir John Denham (1614/15–1669), poet and courtier. It has not been possible to identify the so-called colonel.

151. The Hon. Philip Howard MP was lampooned as Sir Positive At-All in Thomas Shadwell's play *The Sullen Lovers* (1668). Sir Positive is described in the Dramatis Personæ as 'a foolish knight that pretends to understand everything in the world, and will suffer no man to understand anything in his company; so foolishly positive that he will never be convinced of an error, though never so gross.' [see Henning, *The House of Commons 1660–1690*].

152. Oranges were relatively expensive fruits in England at this time.

NOTES

153. The full chorus of this song is 'So old, so wondrous old, in the nonage of Time, | E'er Adam wore a Beard, she was in her Prime.'

154. 'Now, that will finish the Manichees!' exclaimed Thomas Aquinas when dining with Louis IX. He was actually referring to the Cathars, another Christian sect considered heretical.

155. Sir Edward Atkyns (1587–1669). Although a royalist, he was created a Baron of the Exchequer (not a peerage but high judicial office) during the Commonwealth and gave Oliver Cromwell legal advice. Following the Restoration, he was recreated a Baron of the Exchequer and knighted.

156. Sir Edward died in his 83rd year, 1669, the very year in which the *Vindication* was published.

157. A common legal maxim: our reasoning is the same for things which are not seen and those which don't exist.

158. Joan of Arc.

Bibliography

ALLEN, WILLIAM [SILIUS TITUS & EDWARD SEXBY], *Killing, No Murder: With Some Additions: Briefly Discourst in Three Questions, Fit for Public View; to Deter and Prevent Single Persons, and Councils, From Usurping Supream Povver*, London: 1659.

ANON, 'Henry Hills, Official Printer,' *Baptist Quarterly*, (6) 5, January 1932, pp. 215–217.

BAKER, SIR RICHARD, *A Chronicle of the Kings of England From the Time of the Romans Government Unto the Raigne of King Charles: Containing All Passages of State and Church, With All Other Observations Proper for a Chronicle: Faithfully Collected Out of Authours Ancient and Moderne; & Digested Into a New Method*, London: Printed for Daniel Frere, 1643.

BLAGDEN, CYPRIAN, *The Stationers' Company: A History, 1403–1959*, London: Allen & Unwin, 1960.

[Gauden, John (attribute to Charles I)], Εἰκὼν Βασιλική [*Eikōn Basilikē*]. *The Pourtraicture of His Sacred Majestie in his solitudes and sufferings. Together with his Maiesties praiers delivered to Doctor Juxon immediately before his death*, [London: Printed by Henry Hills], 1649.

HENNING, BASIL DUKE, *The House of Commons, 1660–1690*, London: Secker & Warburg, 1983.

Hill, Christopher, *Puritanism and Revolution: Studies in Interpretation of the English Revolution of the 17th Century*, London: Secker & Warburg, 1958.

HOWELL, JAMES, *Londinopolis: An Historicall Discourse Or Perlustration of the City of London, the Imperial Chamber, and Chief Emporium of Great Britain: Whereunto is Added Another of the City of Westminster, With the Courts of Justice, Antiquities, and New Buildings Thereunto Belonging*, London: Printed by J. Streater, for Henry Twiford [etc.], 1657.

KIPPIS, ANDREW, *Biographia Britannica, or, the Lives of the Most Eminent Persons Who Have Flourished in Great Britain and Ireland, from the Earlies Ages, to the Present Times*, vol. I, 2nd edn, London: Printed by W. and A. Strahan for C. Bathurst [etc.], 1778.

L'ESTRANGE, SIR ROGER, *Considerations and Proposals in Order to the*

BIBLIOGRAPHY

Regulation of the Press: Together With Diverse Instances of Treasonous, and Seditious Pamphlets, Proving the Necessity Thereof, London: Printed by A.C., 1663.

MYERS, ROBIN, & MICHAEL HARRIS, *The Stationers' Company and the Book Trade, 1550–1990,* Winchester: St.Paul's Bibliographies, 1997.

NEWMAN, PETER, *Atlas of the English Civil War,* Beckenham: Croom Helm, 1985.

Oxford Dictionary of National Biography [Online Resource], Oxford: Oxford University Press, 2004–.

[PARLIAMENT (COMMONS PROCEEDINGS)], *A Remonstrance of the State of the Kingdom. 15 Decemb. With the Addition of the Humble Remonstrance, and Petition of the Lords and Commons to the Kings Majesty,* London: I. Hunscutt, 1641.

PULTON, FERDINANDO, *An Abstract of All the Penall Statutes Which be Generall in Force and Vse: Wherein is Contayned the Effect of All Those Statutes Which Doe Threaten to the Offendours Therof the Losse of Life, Member, Lands, Goods Or Other Punishment Or Forfeiture Whatsoeuer; Whereunto is Also Added in Theire Apte Titles, the Effect of Such Other Statutes Wherin There is Any Thing Materiall and Most Necessary for Eche Subiect to Know ; Moreouer the Aucthoritie and Duetie of All Justices of Peace, Shirifes, Coroners, Eschetors, Maiors, Baylifes, Customers, Comptroulers of Custome, Stewardes of Leetes and Liberties, Aulnegers and Purueiours and What Things By the Letter of Seuerall Statutes in Force May, Ought, Or Are Compellable to Doe,* [London]: In Ædibus Richardi Tottelli, 1577.

QUEEN MARY, UNIVERSITY OF LONDON, 'A Brief History of the King's Printing House in the Reign of James I', ([n.d.]) <http://www.english.qmul.ac.uk/kingsprinter/publications/transcripts/Reader_Aids/Brief_History.html>.

Richardson, John, *The Annals of London: A Year-By-Year Record of a Thousand Years of History,* London: Cassell, 2000.

ROYLE, TREVOR, *Civil War: The Wars of the Three Kingdoms 1638–1660,* London: Little, Brown, 2004.

RUFINUS OF AQUILEIA, *Incipit Exposicio Sancti Ieronimi in Simbolum Apostoloru[m] Ad Papa[m] Laure[n]Tiu[m],* Oxford: [s.n.], [1478 but bearing the date M.cccc.lxviij].

SHADWELL, THOMAS, *The Sullen Lovers, Or, the Impertinents: A Comedy*

BIBLIOGRAPHY

Acted By His Highness the Duke of Yorkes Servants, [London] In the Savoy: Printed for Henry Herringman [etc.], 1668.

STOW, JOHN, *A Svrvay of London: Contayning the Originall, Antiquity, Increase, Moderne Estate, and Description of That Citie, Written in the Year 1598... Also an Apologie (Or Defence) Against the Opinion of Some Men, Concerning That Citie, the Greatnesse Thereof. With an Appendix, Containing in Latine, Libellum De Situ & Nobilitate Londini: Written By William Fitzstephen, in the Raigne of Henry the Second,* London: imprinted by [John Windet for] Iohn Wolfe..., 1598.

[TWYN, JOHN], *A Treatise of the Execution of Justice: Wherein is Clearly Proved That the Execution of Judgement and Justice is as Well the Peoples as the Magistrates Duty, and That if Magistrates Pervert Judgement, the People Are Bound By the Law of God to Execute Judgement Without Them and Upon Them,* [London?]: [s.n.], [166–].

Index

Acheson, Captain 64
Acheson, Lady Martha 62, 63, 135
  attends masque 67
  estate at Faringdon 68
    Sir Robert Pye's litigation over 69
  estate in the Strand 69
  executors and guardians 65
  falls sick of quinsy 67
  marriage settlement 68
  marries Atkyns 68
  uncle abuses his position as trustee 69
Acheson, Sir Patrick 69, 70, 135
Albemarle ii
  George Monck, 1st Duke of 129
aliquis in omnibus [etc.] 131
Allen, William (pseudonym) 132
  Killing no Murder 27
Allin, Thomas 47, 133
Anne of Denmark 135
Aquinas, Thomas 139
Arundell, Ann Lady 57
  goodness of 61
  kindness to Atkyns 61
Arundell of Trerice, Richard Arundell, 1st Baron 82, 136
Arundell of Wardour, Thomas Arundell, 1st Baron 57, 134
  catholic 58
  Count of Holy Roman Empire 58
  count of Holy Roman Empires 134
  kindness to Atkyns 61
Arundell of Wardour, Thomas Arundell, 2nd Baron 78, 134
  takes a dragoon's colours 79
Arundell, Thomas, jr 57, 58, 134
  dies at Orléans 59
  Dr Labourne tutor to 59
Astley, Sir Jacob 95, 137
Atkyns, Sir Edward 56
  created Baron of the Exchequer 139
  dies 139
  entertains Richard A 120
  kindness to Martha A 117, 121
  tries to reconcile Richard and Martha 118
Atkyns, Martha
  arrested by Sir Jacob Astley's soldiers 95
  entertains dean of Gloucester in her chamber 110–111
  estate in the Strand 116
  finds house full of soldiers 95
  house and grounds ruined during sequestration 99
  husband rescues estate with proceeds of sale of his owm 98
  Middlesex estate freed from sequestration 100
  miscarries 95
  property leased from Church 100
  requests lock of husband's hair 95
  shown kindness by Sir Edward Atkyns 117, 121
  Sir Edward A's kindness to 117
  taken prisoner at Nettlebed 95
  travels to London through parliamentary lines 94
  treated better by Atkyns's uncle than her own 95
  treated with respect by Sir John Cademan 95
  tries to stop Richard renewing lease 124

Atkyns, Richard
  aided by Dr Labourne 101
  at seige of Bristol 72
  borrows money to help uncle save Vyne 69
  breaks arm as a baby 56
  case against Stationers and Flesher 23
  challenged to a duel 96
  charges enemy at Little Dean 72
  commands forelorn hope
    at Reading 74
    three times 72
  dines with Prince Maurice 86
  discovers enemy near Devizes 86
  does not drink between meals 60
  education 56–57
  engaged in masque for Queen 62
  engages heselrige 88–90
  estate sequestered 98
  exposes himself to court 67
  falls ill 67
  fined for cheese fed to grayounds 80
  first debt 63
  forced to befriend Speaker of House of Commons 98
  foreign travels 57–60
  head kicked by horse 56
  hearing for debt at Goldsmiths' Hall 98
  held for debt in Poultry Compter 98
  horse collapses 72
  horse left at Marlborough by groom 91
  liable for wife's debts 116
  loses £300 hatband 67
  marches to relieve Reading 74
  marries Lady Acheson 68
  meets up with Carnarvon's regiment 78
  mother finds an heiress for 62
  musket ball from own side just misses 73
  no longer avoids drinking between meals 61
  Original and Growth
    copy text for this edition v
    fanciful v
    lack of historical accuracy vi
  parentage 55
  pays dancing master to take his part in masque 67
  performs in a masque 62
  performs in another masque 66
  performs masque at Queen's insistence 67
  prefers receiving sacrament to breakfast with Prince Maurice 80
  presents pair of pistols to Prince Maurice 80
  pursues Lady Acheson 65
  raises cavalry troop 71
  relentless snobbery vi
  reminds king of service in royalist cause vi
  rescues wife's estate with proceeds of sale of his own 98
  retires to Oxford after siege of Gloucester 94
  returns to England 60
  royalist sympathies become public knowledge 71
  stays with Sir Edward 120
  stranger in own house 115
  sues Sir Robert Viner 102
  sues Stationers 102
  summons from Prince Maurice 91
  sycophancy vi
  takes Heselrige prisoner 90
  Vindication
    copy text for this edition v

wife tries to stop renewing lease 124
Atkyns, Richard, of Tuffley 55, 134
 justice of Welsh sessions, QC 55
Atkyns & wife v. Flesher & Stationers 23
Bacon, Sir Francis 130
Baker, Sir Richard 130
 Chronicle 12
Balliol College, Oxford 134
Barker, Christopher 25, 131
Barker, Mr
 killed 82
Barker, Robert 131
Barkstead, John 100, 138
Bath, Somerset 77
 parliamentary troops quartered at 81
 royalists enter 92
Bedford, William Russell, 5th Earl 58, 134
Berkshire 136
Bible 131
 printing of in gift of monarch 19
Bill, John 25, 131
book
 prices 27
Book of Common Prayer 131
books
 burning of 18
 licensing of 19
 regulating prices of 48
booksellers
 not essential 22
Bourchier, Thomas vii, 14, 15
Bray, Edmund, 1st Baron 56, 134
Bray, Margaret 134
Bray, Sir John 134
Bray, Sir Reginald 134
Bristol
 royalists retreat to 92

 royalists take 93
 seige of 72
Brydges, Catherine 134
Buck, Col. Brutus 77, 93
Burrell, Col. Robert 82
Cademan, Sir John 137
 treats Mrs Atkyns with respect 95
Calais 58
Cambrai 59
Cambridge
 St John's College v
Carew, Sir Horatio 77
Carnarvon, Robert Dormer, 1st Earl of 76, 77, 92, 94, 136
 Atkyns meets up with 78
 shot in leg while charging enemy 83
Cathar 139
Caversham Bridge
 fighting at 74
Caversham Park 74, 136
Caxton, William vi, 12, 14
Chandos, Edmund Brydges, 2nd Baron 134
Chandos, George Brydges, 6th Baron 71, 72, 135
 uses Atkyns's troop hard 72
Charles I 129
 as art connoisseur 21
 attempt to arrest Heselrige 137
 calls for execution of 132
 Eikōn Basilikē [Εἰκὼν Βασιλική] 135
 picture collection 131
Charles II
 complains of brewers 133
 Loggan engraving ii, 129
Cheddar cheese
 Atkyns fined for cheese fed to his greyhounds 80
Chewton Mendip, Somerset 77, 78
Clarendon, Edward Hyde, First Earl of ii, 129

Cole, Dr
  chaplain to Prince Maurice 80
Colnbrook, Berkshire 65, 135
common law, courts of 129
Common Pleas, court of 130
Common-Prayer, Book of
  printing of in gift of monarch 19
composition (of debt) 137
conciliar courts 129
Cornish Regiment of Foot 76, 85
Corsellis, Frederick vii, 15, 130
Coster, Laurens Janszoon 129, 130
Countess of Pembroke's Arcadia 136
Covent Garden
  Atkins has his head kicked by horse 56
Crawford, Ludovic Lindsay, 16th Earl of 87, 136
Cromwell, Oliver
  assassination of advocated 132
Crookhorn, Hampshire 76, 136
Crypt School, Gloucester 57, 134
Darcy, Edward 47, 133
Dean, forest of
  Atkyns charges enemy in 72
Dean, Forest of 135
Denham, Sir John 113, 138
Devizes, Wiltshire 86, 87, 91
  royalists retreat to 85
Diana, temple of
  under St Paul's 132
divine right of kings vi
divorce
  whether acceptable 122–123
Dorchester
  Royalists take 94
Douai
  English seminary at 59, 134
Dover 58, 60
Dudston 30, 132
Edge Hill, Warwickshire

battle of 71, 135
Edmund 134
Edward IV 13
Eikōn Basilikē [Εἰκὼν Βασιλική] 129, 135
Elder, Seneca the 132
Elizabeth I 28, 131, 134
  grants patent for printing law books 19
  protect royal privilege of printing 19
  visit to Queenborough 132
Essex, Robert Devereux, 3rd Earl of 74, 129, 136
Exchequer, Remembrancer of the 135
Faringdon, Oxfordshire 65
  Atkyns stops at 87
  Lady Acheson's house at 66, 68
  litigation over Acheson estate 69
Finnett, Lady 68
Finnett, Sir John 135
  gives away Lady Acheson 68
Flesher, Miles 24, 25, 131
  Atkyns's case against 23
forelorn hope
  Atkyns commands at Reading 74
  Atkyns commands three times 72
Fortescue, John 46, 133
France
  regulation of printing in 49
Frederick V 135
Gallio, Junius Annaeus 132
Gentilis, Robert 133
gentlemen
  not printers 20
Germany
  regulation of printing in 49
Glastonbury, Somerset
  Atkyns quartered at 80
Gloucester

Gloucester
  area very pro-Parliament 70
  bishop 112
  Crypt School 57, 134
  loses King's Barton and Dudston 30, 132
  merchants attempt to bribe royalists 135
  Parliamentary army marches on 73
  Parliamentary garrison 135
  Royalist siege of 94, 137
Gloucester Cathedral
  dean and chapter 66, 100, 101
  dean entertained in Martha Atkyns's chamber 110–111
  dean opposes renewal of lease to Atkyns 101, 112
  land expropriated by Parliament 100
Goldsmiths' Hall
  Atkyns's hearing for debt 98
  Committee for Compounding 137
Grandison, William Villiers, 2nd Viscount 72, 136
Grand Remonstrance 129
grant of privilege
  for printing law books 37–50
  condemned as monopolistic 39
Greene, Robert 130
Grenville, Sir Bevil 136
  killed 83, 84
Groot, Hugo de 133
Guilford, Elizabeth Boyle, Countess of 138
  charms dean of Gloucester 101
Gutenberg, Johannes vii, 12, 14, 129, 130
Haarlem 14, 129, 130
  Gutenberg's supposed printing works at vii
Hall, Joseph 11, 123, 130
Hamilton, Sir James 76, 84, 86
  commands Atkyns to seek out enemy 86
  injured 91
Hanmer, Captain
  killed in ambush 73
Henly-on-Thames 65
Henrietta Maria
  masque for 62
Henry Col. Norwood 101
Henry V vii
Henry VI 12, 13, 14
Henry VIII
  act controlling cost of books 16
heresy 28
Hertford, William Seymour, 1st Marquess of 82, 136
  regiment 81
Heselrige, Sir Arthur
  Atkyns engages 88–90
  Atkyns takes prisoner 90
  Charles I's attempt to arrest 137
  Cuirassiers ('Lobsters') 137
  rescued 90
Hicks Hall, Clerkenwell 100, 138
Hills, Henry x, 129
Historia della Sacra Inquisitione 133
Holland
  regulation of printing in 49
Holmes, Mr 89
  Atkyns's cornet 79
holy see
  regulation of printing by 49
Hopton, Ralph, 1st Baron 76, 84, 136
  injured in explosion of ammunition wagon 84
Howard, Hon. Philip 138
Howell, James 130
  Londinopolis 12
informers
  rewarding of 33
Ireland
  Acheson property in 70

Irving
- enters Atkyns's service 64
- refuses Parliamentary lieutenancy 71
- relinquishes role in gendarmerie to join Atkyns 71

Islip, John 12, 130

James I
- grants law book patent to More 19

Jewel, John 58, 134
Kellison, Matthew 134
Killing no Murder 27, 132
King's Barton 30, 132
King's Bench, court of the 130
Kings' Bench Prison 98
King's Grant of Privilege 37–51
- copy text v

Kitely, Captain 79
Labourne, Dr
- aids Atkyns's negotiations with dean of Gloucester 101
- civility of to Atkyns and White 59
- Thomas Arundell's tutor 59

Lacedemonians
- child-rearing techniques 27

Lambeth Palace 14, 130
- supposedly holds document supporting Atkyns's history of printing vii

Lambourn, Berkshire 136
- Atkyns has horse reshoed at 87

law books
- letters patent for printing 19, 131

law, common
- grant of privilege to print books on 37–50

law, written & unwritten 46
Lee, Mr
- killed 82

Legge, Col. William 75, 96, 136
Leighton, Major 73
Lenthall, Lady 98

Lenthall, William 138
L'Estrange, Roger 34
- *Considerations and Proposals* 133

libel 28
Lincolns Inn
- Atkyns enters (briefly) 57

Littledean
- Atkyns charges enemy at 72

Loggan, David ii
- engraving of Charles II 129

Londinopolis 12, 130
London 66
Long, Colonel Sir James 87, 136
Long Parliament 50
Long, Sir Robert 79
Louis XIII 134
Lower, Major
- killed 84

Lyme Regis, Dorset
- siege of 94

Magdalen College, Oxford 97
Mainz 12
Marie de' Medici 135
- prepares masque for her daughter 62

Marlborough, Wiltshire
- Atkyns's injured horse left at 91
- Atkyns summoned to 87

marriage
- causes for divorce 121–122

Marshfield, Gloucestershire 84
- royalist troops quartered at 81

Mary, Queen of England
Stationers' Company
- reasons for incorporation 31
- Stationers incorporated in reign of 17

masque 66
- for Henrietta Maria 62

Maurice, Prince 76, 80, 84, 88, 136
- Atkyns joins regiment of 72

[148]

freed 79
injured and taken prisoner 77
invites Atkyns to supper 86
receives pair of pistols from Atkyns 80
regiment 81
summons Atkyns 91
Micheldever, Hampshire
  Acheson property in 70
Molesworth, Lt Col. Guy 91
Monck, George ii, 129
monopolies 44
  illegal when restrain of trade 41
  not all illegal 39
  objection to grants of privilege creating 39
  public benefit can justify 40
  public benefit of some 39
Montmorency, Henri 2nd Duke of 59, 134
More, John 135
  granted law book patent 19, 43, 131
More, Thomas
  Utopia 47, 133
More, William 135
music
  despicable as a trade 18
mystery or art of stationers 131
National Trust 134
Nettlebed, Oxfordshire 74
  Martha Atkins taken prisoner 95
Neville, Lt Col. Richard
  gallantry 85
Neville, Lt Col.Richard 136
Newcastle, William Cavendish, Marquess (later Duke) of 94, 137
Norton, Bonham 131
  patent for law books 19, 43
Norton, Frances 138
Norton, Lady Frances 111

Norton, Sarah 131
Norwood, Colonel Henry 138
Orléans
  Atkyns and Arundell winter in 59
  Gaston, Duke of 59, 134
Oxford 96
  Atkyns retires to after siege of Gloucester 94
  Prince Maurice arrives at 87
  supposedly first place of printing after Mainz and Haarlem vii, 15
Oxford University 57
  Balliol College 134
Palmer, Sir James 21, 131
Paris 58, 59
  uprising of pages in 59
Parliament, Long 50
patent for printing viii, 19, 21, 23, 24, 27, 28, 32, 34, 43
  effective regulation by 22
  law books 24
Patent Office 131
patents
  encourage industry and invention 39–52
  modern concept of 130
patents, royal 130
  source of friction between King and Parliament vi
Paul, Saint 132
Philipson, Anne 134
Pliny the Elder 131
Poole
  royalists take 94
Popish Plot 129
Poultry Compter
  Atkyns imprisoned for debt 98
prerogatives, royal
  extent of 42
presbyterianism 129

Parliament's wish to impose 96
printers
  King's 24, 25, 48
  King's sworn servants 16
  university 49
printing
  abuses of 48
  Acheson patent for 70
  art or mechanic trade? 17
  benefit of 12
  divine nature of 12
  in the king's gift 28
  introduction of into England 12
  no magic 20
  penalty for unlicensed 33
  regulation of in other countries 49
  royal patent 19, 20, 21, 22, 23, 24, 27, 28, 32, 34
    law books 24
  royal prerogative 33
Pulton, Ferdinando
  Abstract 24, 131
Pye, Sir Robert 135
  litigation over Acheson's Faringdon estate 69
Queenborough, Kent 29, 132
Quinsy 135
Reading, Berkshire 136
  Atkyns marches to relieve 74
  Martha Atkyns held in 95
Reading, Jacob Astley, 1st Baron Astley of 137
restraint of trade 41
Richard III
  restraint of aliens 16
Ripple Field 136
Rochester, Henry Wilmot, 1st Earl of 87, 88, 92, 136
Rood, Theoderic 130
Roundway Down
  royalist victory at 90, 137

royal prerogative 41–42, 44
Rufinus of Aquileia vi
  Commentarius in Symbolum Apostolorum 13, 130
rumper 138
Rupert, Prince 75, 93
St Albans
  Henry VI founds press at 15
St Albans 130
St Paul's Cathedral
  on site of temple of Diana 132
Saint-Quentin 59
salt works
  owned by Acheson family 70
Sandys, Elizabeth, Lady 56, 57, 134
Sandys genealogy 134
Sandys, Lieutenant Thomas
  Atkyns finds recovering at Bristol 92
  feels bound by his parole 92
  taken prisoner 79
Sandys, Sir Edwin 134
  Atkyns's grandfather 56
Sandys, Sir William 63, 66
  seeks Atkyns's help in saving Vyne 68
Sandys, William, 1st Baron 134
Sandys, William 3rd Baron 56, 134
sedition 29
  encouraged by Stationers' Company 32
Seneca the Elder 132
Servita, Paolo 44, 133
Sexby, Edward 132
  Killing no Murder 27
Shadwell, Thomas
  The Sullen Lovers 138
Sheldon, Gilbert ii, 129
Sheldon, Major Thomas 73, 80, 84, 136
  dies 85

Sidney, Sir Philip
  Countess of Pembroke's Arcadia 136
Slanning, Sir Nicholas 93, 137
Smith, Major Paul 88
Solemn League and Covenant 129
Spartans 132
  child-rearing techniques 27
Speaker of House of Commons
  Atkyns forced to make friends with 98
Star Chamber 28, 129
Stationers' Company
  Atkyns calls for powers to be limited 33
  Atkyns's case against 23, 102
  charter of 29
  corrupt 31
  encourages printing of sedition and treason 32
  exceeds its charter 32
  fails to control press 29
  fails to fulfil purpose 31
  Hall Book 24
  incorporation (1557) 16, 131
  letters patent 1603 131
  mystery of 18, 131
  not fit to curb evils of press vii, 26
  not to be trusted 30
  obtain control of printing by deceit 28
  opposes royal patentees 31, 32
  promotes deceit in book-trade 21
Stationers' Hall 34
  near temple of Diana 132
Statute of Monopolies 130
Stow, John vi, 12, 13, 130
Strafford, Thomas Wentworth, 1st Earl of 135
  Bill of Attainder 135, 137
  grants creditors possession of Acheson's Irish property 70
  trial of 70
Strand, the
  Lady Acheson's estate in 69
Survey of London 12, 130
Taunton Deane, Somerset 76, 136
Tewkesbury
  Parliamentarians surprise Royalists at 73
Thomas Aquinas 114
Titus, Silius 132
  Killing no Murder 27
Tog Hill, Gloucestershire 84
  battle of 81
Tottel, Richard
  patent for law books 19, 131
    granted by Elizabeth I 43
    renewed by Elizabeth I 43
  treason 29, 30
    encouraged by Stationers' Company 32
Trevanion, Col. John 93, 137
Tuffley 134
  Richard Atkyns of 55
Turner, William 14, 130
Twyn, John
  A Treatise of the Execution of Justice 132
  convicted of treason 30, 132
university printers 49
unlicensed publications 29
Uxbridge, Treaty at 95, 137
Vantruske, Major 82
Venetian ambassador
  chastises wife 67
Viner, Sir Robert 102, 138
Vyne, the, Hampshire 62, 134
  home of Atkyns's grandmother 57
Wales
  R. Atkyns of Tuffley justice in 55
Waller, Sir William 72, 76, 77, 136, 137

Wallingford
  humanity of 92
Wallingford 74
Wall, Lieutenant Colonel
  killed 84
Walsh, Sir Robert 83, 92, 136
Washnage, Cornet 84
  killed by explosion 85
Wells, Somerset 77
  Atkyns arrives at 79
Western Association army 137
Westminster
  Henry VI founds press at 15
Westminster Abbey
  printing at 12
White, Master
  accompanies Atkyns and Arundell to Europe 58
Wight, Thomas 131
  patent for law books 19, 43
Windsor Great Park
  enclosure of 27
Wren, Sir Christopher 132
Wroth, Sir Henry 137
Yestweirt, Nicasius
  patent for law books 19, 43, 131

TIGER OF THE STRIPE

Typeset in Jannon Text Moderne
and Jannon Text Moderne Swash
(now superseded by Jannon 10 Pro)
from the Storm Type Foundry.
The blackletter is English Textura
from Hoefler & Frere-Jones
and the Greek is
Adobe Garamond Premier Pro.
The cover is set in
IM Fell French Canon Pro
and Jannon.

www.ingramcontent.com/pod-product-compliance
Lightning Source LLC
LaVergne TN
LVHW091551070426
835507LV00010B/803